Springer Proceedings in Mathematics & Statistics

Volume 531

This book series features volumes composed of selected contributions from workshops and conferences in all areas of current research in mathematics and statistics, including data science, operations research and optimization. In addition to an overall evaluation of the interest, scientific quality, and timeliness of each proposal at the hands of the publisher, individual contributions are all refereed to the high quality standards of leading journals in the field. Thus, this series provides the research community with well-edited, authoritative reports on developments in the most exciting areas of mathematical and statistical research today.

Sandra De Iaco • Monica Palma • Donato Posa
Editors

Exploration of Spatio-Temporal Environmental Conditions: Harmonized Databases and Analytical Techniques

ECoST-DATA, Bari, Italy, July 3-4, 2025

Springer

Editors
Sandra De Iaco
University of Salento
Lecce, Italy

Monica Palma
University of Salento
Lecce, Italy

Donato Posa
Unversity of Salento
Lecce, Italy

Workshop on Exploration of Spatio-Temporal Environmental Conditions ecost-data ecost-data2025 ECoST-DATA Bari, Italy 2025 07 03 2025 07 04 https://sites.google.com/unisalento.it/ecost-dataproject/workshop

ISSN 2194-1009 ISSN 2194-1017 (electronic)
Springer Proceedings in Mathematics & Statistics
ISBN 978-3-032-17525-0 ISBN 978-3-032-17526-7 (eBook)
https://doi.org/10.1007/978-3-032-17526-7

Mathematics Subject Classification: 60, 62

This work was supported by University of Salento.

This Springer imprint is published by the registered company Springer Nature Switzerland AG
The registered company address is: Gewerbestrasse 11, 6330 Cham, Switzerland

Preface

This volume collects the scientific contributions presented at the Workshop "Exploration of Spatio-Temporal Environmental Conditions: Harmonized Databases and Analytical Techniques (ECoST-DATA)", held at the University of Bari on July 3-4, 2025, as Satellite event of DUST 2025. It is a key result of the ECoST-DATA Project, funded by European Union – NextGenerationEU with the Cascade Open Calls published by ALMA MATER STUDIORUM – University of Bologna, inside the Project GRINS within the PNRR – Mission 4, Component 2, Investment 1.3 "Partnership extended to Universities, Research Centers, Firms and research projects funding", D.D. 341 of 15/03/2022, CUP: J33C22002910001.

The contents are dedicated to advanced tools and techniques for analyzing and predicting environmental variables with a spatial or spatio-temporal structure. In particular, they include theoretical reviews on spatio-temporal covariance modelling as well as innovative approaches for assessing environmental quality and its effects on climate change by integrating georeferenced data from multiple sources.

Among the primary objectives of this collection, it is worth mentioning: transfer advanced analytic skills for the modelling, estimation, and prediction of spatial and spatio-temporal environmental variables; provide practical guidance on integrating heterogeneous georeferenced data while accounting for measurement error and spatial misalignment; support risk assessment by equipping practitioners with methods to quantify exposure and uncertainty relevant to human health; and inform sustainability planning through models that identify spatially and temporally targeted mitigation or adaptation strategies.

As a result, the contributions are intended for researchers and graduate students in spatial statistics, environmental science, and geosciences; data scientists and practitioners working on environmental monitoring, climate impacts, and public-health risk assessment; policy analysts interested in data-driven sustainability planning.

Regarding the outcomes, readers will gain both the theoretical grounding and the practical toolset necessary to build, validate, and deploy spatio-temporal models for environmental variables, produce reliable forecasts, and translate model outputs into actionable insights for health and sustainability interventions.

The volume is organized in three parts.

In the first one, *Random fields and Covariance modelling*, Posa clarifies the relationship between covariance functions and positive definite functions, emphasizing that Crum's theorem is the necessary generalization of Bochner's theorem. De Iaco discusses on some families of spatio-temporal complex covariance functions for vectorial data. It is also described how these functions can be generalized to the spatio-temporal setting, as well as how some other complex-valued covariance families can be constructed as positive mixtures of infinite components or as a convolution of the real part. Hristopulos presents the hybrid spectral approach for generating physically inspired, non-separable covariance kernels from stochastic ordinary differential equations. Pere et al. propose the stationary subspace analysis to decompose a random field into independent stationary and non-stationary subspaces.

In Part II, *Environmental control and integration*, Stein focuses on the use of modern deep learning methods for the analysis and management of large data volumes. His study explores the use of satellite imagery and time series to analyze air quality and monitor environmental changes. A further challenging approach is presented in Gómez-Hernández et al. to identify atmospheric dust sources, using the ensemble smoother with multiple data assimilation methodology. Capasso discusses the optimization of pollution control within an economic growth system, presenting a spatially structured model. This model analyzes an optimal control problem designed to maximize intertemporal utility while minimizing pollution, the disutility of taxation, and intervention costs. Additionally, the study by Duttilo and Di Battista explores the functional structure and variation of Italian carbon emissions from electricity using functional data analysis. Daily and monthly profiles are constructed using Fourier basis functions and analyzed at national and electricity market zone levels.

The volume concludes with Part III, *Environmental Analysis*, where Jalilian et al. apply a flexible generalized additive model framework for surface ozone prediction, using spatio-temporally varying coefficients to correct biases between satellite and in situ data. Gomes et al. investigate the geostatistical characterization of climate extremes dynamics, focusing on drought. A Random Forest-based short-term alert system is presented for predicting drought severity classes at a monthly scale, demonstrating strong predictive performance for local areas and offering a flexible tool for managing drought-related risks. For environmental quality assessment, the work by Masoumi et al. addresses a gap in biodiversity modelling in the Apulia region by comparing the performance of various machine learning models (Random Forests, Support Vector Machines, Extreme Gradient Boosting, Decision Trees, and Linear Regression) for assessing the effects of environmental variables on the multilevel biodiversity index. Furthermore, in Cappello et al., spatio-temporal blind source separation model is reviewed and employed to model and predict ground-level ozone concentration in a complex multivariate spatio-temporal framework.

Lecce, Italy
September 2025

Sandra De Iaco
Donato Posa
Monica Palma

Acknowledgments This research was supported by the European Union-NextGenerationEU with the Cascade Open Calls published by ALMA MATER STUDIORUM - University of Bologna, inside the Project GRINS funded by PNRR - Mission 4, Component 2, Investment 1.3 "Partnership extended to Universities, Research Centers, Firms and research projects funding", D.D. 341 of 15/03/2022, "ECoST-DATA, Exploring Spatio-Temporal Environmental Conditions: Harmonized Databases and Analytical Techniques", CUP: J33C22002910001.

Contents

Part III Environmental Analysis

Part I
Random Fields and Covariance Modelling

Covariance Functions

Donato Posa

Abstract This chapter focuses on Bochner's theorem and its connection to positive definite functions and covariance functions. There exists a common misunderstanding among scientists regarding the precise relationships between these mathematical concepts. A key point is the limited scope of Bochner's theorem, as it only applies to continuous covariance functions. It will be emphasized the practical need for a generalization of the theorem, particularly for applications involving temporal, spatial, and spatiotemporal data analysis, where correlation models often exhibit discontinuity at the origin. Ultimately, the chapter aims to clarify and expand upon the theoretical understanding of these functions.

Keywords Covariance functions · Positive definite functions · Bochner's theorem

1 Introduction

Various areas of mathematics (Fourier analysis, complex analysis and probability, just to mention some of them) have to deal with the family of functions which are positive definite (*PD*) [29]. This last family of functions [20, 28] matches with the family of covariance functions: this relevant remark has been very often neglected in the literature or very frequently confused with the famous Bochner's theorem.

There is no need to recall any assumption of regularity or continuity to define the family of *PD* functions. Moreover, the subset of the continuous covariance functions is fully portrayed by the theorem of Bochner [2]. For this purpose, it will be underlined that any complex covariance function (*CF*), or equivalently any complex *PD* function, cannot be obtained from Bochner's theorem.

D. Posa (✉)
Department of Economic Sciences, University of Salento, Lecce, Italy
e-mail: donato.posa@unisalento.it

S. De Iaco et al. (eds.), *Exploration of Spatio-Temporal Environmental Conditions: Harmonized Databases and Analytical Techniques*, Springer Proceedings in Mathematics & Statistics 531, https://doi.org/10.1007/978-3-032-17526-7_1

The chapter clarifies the often confused relationships among *PD* functions, strictly positive definite (*SPD*) functions [21, 22] and *CF*s, noting that *CF*s are equivalent to *PD* functions. It highlights that Bochner's theorem only describes the subset of continuous *CF*s, while many real-world applications require models that are discontinuous at the origin. Crum's theorem will be introduced as a crucial generalization that accommodates these discontinuous functions, providing a more comprehensive understanding of correlation theory essential for temporal, spatial and spatiotemporal data analysis. The chapter supplies definitions and properties for each function type, emphasizing the distinctions and implications for various statistical applications.

Apart from the last remark which is essentially related to interpolation problems, the results given hereafter are not restricted to the applications which involve spatial and spatiotemporal analysis, in which several relevant contributions can be found in [5, 7, 9–11, 17, 19]; on the other hand, the same results regard the whole correlation theory and some properties, as well as, in general, all the fields related to *PD* functions.

The chapter outlines key aspects of *CF*s, particularly their properties as *PD* and *SPD* functions, independent of continuity assumptions. It discusses Bochner's theorem and its extensions, highlighting the connection between these function families. It is remarkable to note that Bochner's theorem is not applicable to certain *CF* models, specifically those exhibiting a discontinuity at the origin (known as *nugget effect*) or those that are almost everywhere zero, which are often encountered in time series and spatial analysis.

The present analysis just concerns the univariate case; some useful details regarding the multivariate case can be found in [15] and [30].

2 *CF*s, *PD* and *SPD* Functions

A brief overview on *PD* and *SPD* functions is given hereafter.

Consider the set $\mathbb{C}$ of complex numbers; then

$$C : \mathbb{R}^m \longrightarrow \mathbb{C}$$

is *PD* if, whatever they are $\mathbf{t}_1, \mathbf{t}_2, \ldots, \mathbf{t}_n \in \mathbb{R}^m$ and any choice of $\lambda_k \in \mathbb{C}, k = 1, 2, \ldots, n, \ \forall n \in \mathbb{N}_+$,

$$\sum_{k=1}^{n} \sum_{l=1}^{n} \lambda_k \overline{\lambda}_l C(\mathbf{t}_k - \mathbf{t}_l) \geq 0; \tag{1}$$

moreover, C is *SPD* on $\mathbb{R}^m$ if

$$\sum_{k=1}^{n} \sum_{l=1}^{n} \lambda_k \overline{\lambda}_l C(\mathbf{t}_k - \mathbf{t}_l) > 0, \tag{2}$$

for $\mathbf{t}_1, \mathbf{t}_2, \ldots, \mathbf{t}_n, \mathbf{t}_k \neq \mathbf{t}_l, k \neq l, \ \forall n \in \mathbb{N}_+$ and $\lambda_k \in \mathbb{C}, k = 1, 2, \ldots, n$, with at least one of them different from zero.

The symbol $\mathcal{P}$ will be utilized to represent the family of *PD* functions, whereas the symbol $\mathcal{P}^+$ will be utilized for the family of *SPD* functions, as a consequence $\mathcal{P}^+ \subset \mathcal{P}$.

It is required that a *CF* belongs to the class $\mathcal{P}^+$ to guarantee the invertibility of the kriging matrix for prediction.

2.1 PD Functions and Their Properties

In the following a brief overview concerning the properties of the family $\mathcal{P}$ of *PD* functions defined in (1) is given. Note that none of the hypotheses regarding continuity or regularity is needed.

1. $0 \leq C(\mathbf{0})$.
2. $C(\mathbf{t}) = \overline{C}(-\mathbf{t}), \quad \forall \mathbf{t} \in \mathbb{R}^m$.
3. $C(\mathbf{0}) \geq |C(\mathbf{t})|, \quad \forall \mathbf{t} \in \mathbb{R}^m$.
4. If $C_i, i = 1 \ldots, n$, are *PD* on $\mathbb{R}^m$, then their product is *PD* on $\mathbb{R}^m$.
5. If $C_k, k = 1, \ldots, n$, are *PD* on $\mathbb{R}^m$ with $a_k \geq 0, k = 1, \ldots, n$, then $\sum_{k=1}^{n} a_k C_k(\mathbf{t})$ is *PD* on $\mathbb{R}^m$.
6. If C is *PD*, then C_{re}, i.e., the real component of C, is *PD*.
7. If $C_k, k \in \mathbb{N}$, are *PD* on $\mathbb{R}^m$, as a consequence
$$C(\mathbf{t}) = \lim_{i \to \infty} C_i(\mathbf{t})$$
is *PD* on $\mathbb{R}^m$, as long as the limit exists.
8. If $C(\cdot; \lambda)$ is *PD* on $\mathbb{R}^m$ $\forall \lambda \in A \subseteq \mathbb{R}$, and $\sigma(d\lambda)$ is a measure on A, then
$$\int_A C(\mathbf{t}; \lambda)\sigma(d\lambda),$$
is *PD* on $\mathbb{R}^m$, provided that the integral exists $\forall \mathbf{t} \in \mathbb{R}^m$.
9. Suppose that a function be *PD* on $\mathbb{R}^n$; however nothing can be said about its positive definiteness on $\mathbb{R}^m, n < m$.
10. If C is *PD*, then for all $\mathbf{t}_1, \mathbf{t}_2 \in \mathbb{R}^m$
$$|C(\mathbf{t}_2) - C(\mathbf{t}_1)|^2 \leq 2C(\mathbf{0})Re[C(\mathbf{0}) - C(\mathbf{t}_2 - \mathbf{t}_1)],$$
where $Re[C(\mathbf{0}) - C(\mathbf{t}_2 - \mathbf{t}_1)]$ is the real component of $[C(\mathbf{0}) - C(\mathbf{t}_2 - \mathbf{t}_1)]$.

This last result is particularly useful since it distinguishes the class of the continuous *CF*s.

2.2 SPD Functions and Their Properties

A brief overview concerning the properties for the family $\mathcal{P}^+$ of *SPD* functions is given: even in this case, none of the hypotheses of continuity or regularity is needed.

1. If the function C is *SPD*, then $C(\mathbf{0}) > 0$; on the other hand, vice versa is not true. Indeed, there can be functions such that $C(\mathbf{0}) > 0$, however the same functions are not *SPD*; the cosine *CF* is a typical example.
2. Consider n functions which are *SPD* on the same space $\mathbb{R}^m$; then their sum, as well as their product, is a *SPD* function on the same space.
3. If C is *SPD* on $\mathbb{R}^m$, it could not be *SPD* on $\mathbb{R}^n$, with $n > m$.
4. The real component of a complex *SPD* is always *SPD*.
5. Consider n *PD* functions on $\mathbb{R}^m$. Suppose that there exists $C_k, k \in \{1, \ldots, n\}$, which is *SPD*, with $\alpha_k > 0$, as a consequence
$$C(\mathbf{t}) = \sum_{i=1}^{n} \alpha_i C_i(\mathbf{t})$$
with $\alpha_i \geq 0, i = 1, \ldots, n$, is *SPD* on $\mathbb{R}^m$.
6. If C is *SPD* on $\mathbb{R}^m$, then $C(\mathbf{t}) \neq C(\mathbf{0}), \forall \mathbf{t} \neq \mathbf{0}$.
7. Furthermore, if C is *PD* and there exists $\mathbf{t}$ such that $C(\mathbf{t}) = C(\mathbf{0})$, with $C(\mathbf{0}) > 0$, it comes out that C is not *SPD*.
8. Let $C_1, \ldots, C_n$ be not identically zero *PD* on $\mathbb{R}^m$ [8]. Then,
$$C(\mathbf{t}) = \prod_{i=1}^{n} C_i(\mathbf{t})$$
is *SPD* on $\mathbb{R}^m$ if $\exists i \in \{1, \ldots, n\}$ such that C_i is *SPD*.

Consider the *nugget effect* function [18]

$$C(\mathbf{t}) = \begin{cases} C_0 & \mathbf{t} = \mathbf{0} \\ 0 & \mathbf{t} \neq \mathbf{0}, \end{cases} \tag{3}$$

with $\mathbf{t} \in \mathbb{R}^m$, $C_0 > 0$; this last function presents a discontinuity at the origin and it can be easily proved that it is *SPD*.

2.3 CFs

Consider the complex random function $Z : \mathbb{R}^m \longrightarrow \mathbb{C}$:

$$Z(\mathbf{w}) = Z_1(\mathbf{w}) + i Z_2(\mathbf{w}), \quad \mathbf{w} \in \mathbb{R}^m, \tag{4}$$

with Z_1 and Z_2 real random functions and suppose that (4) is stationary of the second order; as a consequence, the corresponding *CF* is defined as follows:

$$C(\mathbf{t}) = E[(Z(\mathbf{w}) - m)\overline{(Z(\mathbf{w}+\mathbf{t}) - m)}],$$

with $\mathbf{w} \in \mathbb{R}^m$ and $(\mathbf{w}+\mathbf{t}) \in \mathbb{R}^m$, m is the expected value of (4), supposed to be constant. Equivalently, the *CF* can be represented as follows:

$$C(\mathbf{t}) = C_{re}(\mathbf{t}) + iC_{im}(\mathbf{t}),$$

where the sum of C_{Z_1} and C_{Z_2} (the *CF*s of Z_1 and Z_2, respectively) corresponds to the real component C_{re}, whereas the difference of the cross-covariances $C_{Z_2Z_1}$ and $C_{Z_1Z_2}$ is the imaginary part, i.e.,

$$C_{re}(\mathbf{t}) = C_{Z_1}(\mathbf{t}) + C_{Z_2}(\mathbf{t}), \qquad C_{im}(\mathbf{t}) = C_{Z_2Z_1}(\mathbf{t}) - C_{Z_1Z_2}(\mathbf{t}). \tag{5}$$

Note that C_{re} is a real *CF*; on the other hand C_{im} cannot be a *CF*, and this last satisfies the property related to cross-*CF*s, i.e.,

$$C_{Z_1Z_2}(-\mathbf{t}) = C_{Z_2Z_1}(\mathbf{t}).$$

The symbol $\mathcal{C}$ will be used to define the family of *CF*s.

Remarks

- It can be easily established that the class of *CF*s coincides with the class of *PD* functions [25], hence $\mathcal{P} \equiv \mathcal{C}$.
- Taking into account the previous item, the above two classes follow the same properties already established in the previous section. A discussion on some characteristics can also be found in [14, 16].
- By definition, a covariance is a complex function, although several case studies utilize the subclass of the real *CF*s.

3 Bochner's Theorem and the Corresponding Generalization

Theorem 1 *[Bochner [2]] A function ϕ on $\mathbb{R}^m$ is candidate to be a continuous CF if and only if it can be expressed as follows:*

$$\phi(\mathbf{t}) = \int_{\mathbb{R}^m} \exp(i\mathbf{t}^T\boldsymbol{\tau})dF(\boldsymbol{\tau}), \tag{6}$$

where the measure F is bounded and it is usually called spectral distribution function.

In the peculiar case the distribution function F is absolutely continuous; then there exists a nonnegative function f which allows to formulate (6) as follows:

$$\phi(\mathbf{t}) = \int_{\mathbb{R}^m} \exp(i\mathbf{t}^T\boldsymbol{\tau}) f(\boldsymbol{\tau}) d\boldsymbol{\tau}, \tag{7}$$

where $f \in L^1(\mathbb{R}^m)$. The function f is named *spectral density* and it is not necessarily a continuous function; however, ϕ satisfies the following asymptotic condition:

$$\lim_{\|\mathbf{t}\| \to \infty} \phi(\mathbf{t}) = 0. \tag{8}$$

On the other hand, the function ϕ could not be integrable; if ϕ is integrable, then

$$f(\boldsymbol{\tau}) = \frac{1}{(2\pi)^n} \int_{\mathbb{R}^m} \exp(-i\boldsymbol{\tau}^T\mathbf{t})\phi(\mathbf{t}) d\mathbf{t}, \tag{9}$$

in addition f is continuous.

From (9) it follows that if the Fourier transform of a function ϕ is nonnegative, then the same function ϕ is the *CF* of a random field.

The symbol $\mathcal{B}$ will be utilized to denote the class of the continuous *CF*s, already defined through expression (1); as a consequence, $\mathcal{B} \subset \mathcal{P}$. On the other hand, the set of the family of *PD* functions contains the set of the family of functions that are *SPD* ($\mathcal{P}^+ \subset \mathcal{P}$); however, the family of functions which are *SPD* cannot be a subset of $\mathcal{B}$.

Remark The chapter discusses the nature of *CF*s, explaining that they can be either complex or real. This distinction hinges on the properties of the function F. Specifically, if F is asymmetric around the origin or the spectral density f is not an even function, the resulting *CF* will be complex. Conversely, a symmetric F, or an even spectral density f, will yield a real *CF*. The chapter aims to provide a thorough characterization for these continuous *CF*s, building upon prior work.

Corollary 1 *A complex covariance is a continuous function if and only if its real component is continuous at the origin.*

From the previous corollary, it comes out that any *CF* can be represented through relation (6) if and only if it is continuous at the origin. As a consequence, the entire class of *PD* functions (or *CF*s) can be split into two sets, which are disjoint: the family of continuous *PD* (or continuous *CF*s), which can be expressed by Bochner's theorem, and the family of *PD*s (or *CF*s) which present a discontinuity at the origin. Several details concerning the techniques to construct extensive family of continuous complex *CF*s can be found in [23].

It is not ensured that the difference between *CF*s can be an admissible covariance: details to obtain admissible *CF*s starting from such a difference have been discussed in [24].

3.1 SPD Functions and Bochner's Theorem

In the present section some results concerning the class $\mathcal{B}$ of continuous *CF*s, which satisfy Bochner's theorem, will be provided.

Some peculiar properties stemming from Bochner's theorem and concerning *SPD* functions are detailed hereafter.

1. Let f be integrable, nonnegative, continuous and not identically zero; then

$$\phi(\mathbf{t}) = \int_{\mathbb{R}^m} \exp(i\mathbf{t}^T \boldsymbol{\tau}) f(\boldsymbol{\tau}) d\boldsymbol{\tau} \tag{10}$$

   is *SPD* and continuous on $\mathbb{R}^m$.
2. Consider n *CF*s $\phi_1, \ldots, \phi_n$ which are continuous on $\mathbb{R}^m$. Then

$$\phi(\mathbf{t}) = \sum_{i=1}^{n} \alpha_i \phi_i(\mathbf{t}),$$

   with $\alpha_i \geq 0, i = 1, \ldots, n$, is *SPD* and continuous on $\mathbb{R}^m$, if and only if there exists $k \in \{1, \ldots, n\}$ with ϕ_k *SPD* on $\mathbb{R}^m$, $\alpha_k > 0$ [8, 12, 13].
3. Suppose that the support of the measure F in (6) contains an open subset; then ϕ is *SPD* and continuous.

3.2 Crum's Theorem: A Generalization of Bochner's Result

Taking into account all the results of the previous sections, it comes out that any *PD* function (or equivalently any *CF*) cannot be expressed through Bochner's theorem. Unfortunately, this relevant aspect has generated a lot of misunderstanding, as well as confusion, as can be detected through some statements collected from important books on the subject.

1. Statistics for spatial data [4, p. 84]
2. The geometry of random fields [1, p. 25]
3. Correlation theory of stationary and related random functions [31, p. 93]
4. Geostatistics: modeling spatial uncertainty [3, p. 64]

The above sentences underline a common misconception regarding *CF*s and Bochner's theorem: there appears to be a direct correlation between *PD* functions and continuous *CF*s that adhere to Bochner's theorem; this is not universally true. Indeed, the entire category of *CF*s exhibiting discontinuity at the origin cannot be represented by the conditions of Bochner's theorem. A key illustration of this aspect is the nugget effect, which represents a typical example of such a discontinuous function. Hence, it is relevant to differentiate the types of *CF*s, highlighting those that fall outside the scope of traditional Bochner's theorem applications due to their inherent discontinuities.

A first generalization of Bochner's theorem was given by Riesz [27]; a further upgrade was obtained by Crum [6], through the result given hereafter.

Theorem 2 (Crum's Theorem) *A measurable and PD complex function C, defined on $\mathbb{R}^m$, can always be written as follows:*

$$C(\mathbf{t}) = \phi(\mathbf{t}) + \psi(\mathbf{t}), \tag{11}$$

where

$$\phi(\mathbf{t}) = \int_{\mathbb{R}^m} \exp(i\mathbf{t}^T\boldsymbol{\tau})dF(\boldsymbol{\tau}),$$

and ψ is PD; moreover $\psi = 0$ almost everywhere.

In the special case $C(\mathbf{t}) = C(\|\mathbf{t}\|)$ (i.e., C is isotropic) and $m > 1$; then $\psi(\mathbf{t}) = 0$, for $\mathbf{t} \neq \mathbf{0}$.

Unfortunately, the previous result is rarely referred to in the literature, although it is widely utilized in most of the case studies, as underlined hereafter.

In Fig. 1 the relationship among *CF*s, *PD* and *SPD* functions has been properly represented.

3.3 Crum's Theorem in the Applications

It is well acknowledged the importance of the theory of correlation in any context regarding temporal and spatial data analysis, as summarized hereafter.

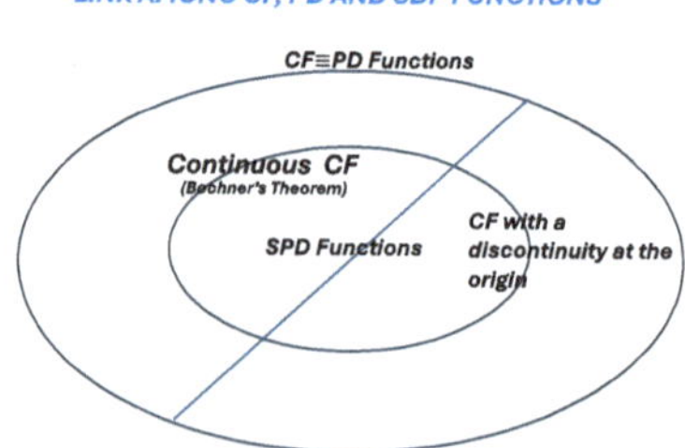

Fig. 1 Relationship among *CF*s, *PD* and *SPD* functions. The set of *CF*s coincides with the set of *PD* functions. The whole family of the *CF* can be partitioned into two subsets: the subset of the continuous *CF* and the subset of the *CF* which presents a discontinuity at the origin. The subset of *SPD* functions is not a subset of the family of continuous *CF*s, neither of the family of the *CF*s which are not continuous at the origin

- In discrete time series analysis [26] stationary models take up a dominant role. At this purpose, the autocorrelation function related to autoregressive, moving average, as well as autoregressive moving average models is zero almost everywhere. In this case, expression (11) shrinks as follows:

$$C(\mathbf{t}) = \psi(\mathbf{t}).$$

For the following autoregressive process:

$$z_t = \varphi_1 z_{t-1} + a_t, \quad -1 < \varphi_1 < 1,$$

the autocorrelation function

$$\rho(k) = \varphi^k, \quad k = 0, 1, 2 \ldots,$$

is nil almost everywhere; then because of (11), $\phi = 0$ and $\psi \equiv \rho$.
- In geostatistics, expression (11) is usually employed; in other terms, the resulting models consist of a linear combination of a nugget effect, described by the function ψ, with a continuous correlation model, described by the function ϕ. On the other hand, Bochner's theorem can be invoked in all the situations corresponding to correlation models which are continuous at the origin.

4 Conclusions

This chapter clarifies the interconnectedness of *PD* functions, *CF*s and continuous *CF*s, addressing prior confusion in the literature. It highlights how Crum's result expands upon Bochner's theorem, further illuminating the relationship between *PD* functions (which are equivalent to *CF*s) and continuous *CF*s. It has been also emphasized the significance of *SPD* functions. Essentially, the work aims to precisely define and distinguish these related mathematical concepts, which are crucial in various analytical applications.

Bibliography

1. Adler, R.J.: The Geometry of Random Fields. John Wiley & Sons, London (1981)
2. Bochner, S.: Lectures on Fourier Integrals. Princeton University Press, Princeton (1959)
3. Chilès, J., Delfiner, P.: Geostatistics: Modeling Spatial Uncertainty. John Wiley & Sons, Hoboken (1999)
4. Cressie, N.: Statistics for Spatial Data. John Wiley & Sons, New York (1993)
5. Cressie, N., Huang, H.: Classes of nonseparable, spatio-temporal stationary covariance functions. J. Am. Stat. Assoc. **94**(448), 1330–1340 (1999)
6. Crum, M.M.: On positive definite functions. Proc. Lond. Math. Soc. **3**(6), 548–560 (1956)

7. De Iaco, S., Posa, D.: Predicting spatio-temporal random fields: some computational aspects. Comput. Geosc. **41**, 12–24 (2012)
8. De Iaco, S., Posa, D.: Strict positive definiteness in geostatistics. Stochastic Environ. Res. Risk Assess. **32**, 577–590 (2018)
9. De Iaco, S., Myers, D.E., Posa, D.: Space-time analysis using a general product-sum model. Stat. Probab. Lett. **52**(1), 21–28 (2001)
10. De Iaco, S., Myers, D., Posa, D.: Nonseparable space-time covariance models: some parametric families. Math. Geol. **34**(1), 23–42 (2002)
11. De Iaco, S., Myers, D., Posa, D.: Space-time variograms and a functional form for total air pollution measurements. Comput. Stat. Data Anal. **41**(2), 311–328 (2002)
12. De Iaco, S., Myers, D.E., Posa, D.: Strict positive definiteness of a product of covariance functions. Commun. Stat. Theory Methods **40**(24), 4400–4408 (2011)
13. De Iaco, S., Myers, D., Posa, D.: On strict positive definiteness of product and product-sum covariance models. J. Stat. Plan. Infer. **141**, 1132–1140 (2011)
14. De Iaco, S., Posa, D., Myers, D.: Characteristics of some classes of space-time covariance functions. J. Stat. Plan. Infer. **143**(11), 2002–2015 (2013)
15. De Iaco, S., Myers, D., Palma, M., Posa, D.: Using simultaneous diagonalization to identify a space-time linear coregionalization model. Math. Geosci. **45**(1), 69–86 (2013)
16. De Iaco, S., Posa, D., Cappello, C., Maggio, S.: Isotropy, symmetry, separability and strict positive definiteness for covariance functions: a critical review. Spatial Stat. **29**, 89–108 (2019)
17. Gneiting, T.: Nonseparable, stationary covariance functions for space-time data. J. Am. Stat. Assoc. **97**(458), 590–600 (2002)
18. Journel, A.G., Huijbregts, C.J.: Mining Geostatistics. Academic Press, London (1981)
19. Ma, C.: Families of spatio-temporal stationary covariance models. J. Stat. Plan. Infer. **116**(2), 489–501 (2003)
20. Mathias, M.: Über positive Fourier-integrale. Mathematische Zeitschrift **16**, 103–125 (1923)
21. Menegatto, V.A., Oliveira, C.P., Peron, A.P.: Strictly positive definite Kernels on subsets of the complex plane. Comput. Math. Appl. **51**(8), 1233–1250 (2006)
22. Pinkus, A.: Strictly positive definite functions on a real inner product space. Adv. Comput. Math. **20**(4), 263–271 (2004)
23. Posa, D.: Parametric families for complex valued covariance functions: some results, an overview and critical aspects. Spatial Stat. **39**, 100473 (2020)
24. Posa, D.: Models for the difference of continuous covariance functions. Stochastic Environ. Res. Risk Assess. **35**, 1369–1386 (2021)
25. Posa, D.: Positive definite functions, stationary covariance functions, and Bochner's theorem: some results and a critical overview. Commun. Stat. Theory Methods **53**(15), 5612–5628 (2024)
26. Priestley, M.B.: Spectral Analysis and Time Series. Academic Press, Cambridge (1981)
27. Riesz, F.: Über sätze von Stone und Bochner. Acta Szeged **6**, 184–198 (1933)
28. Schoenberg, I.J.: Metric spaces and completely monotone functions. Ann. Math. **39**(4), 811–841 (1938)
29. Stewart, J.: Positive definite functions and generalizations, an historical survey. Rocky Mt. J. Math. **6**(3), 409–434 (1976)
30. Wackernagel, H.: Multivariate Geostatistics: an introduction with applications. Springer series in Statistics, Springer, Berlin (2003)
31. Yaglom, A.M.: Correlation theory of stationary and related random functions. Springer Series in Statistics, vols. I, II. Springer, Berlin (1987)

cc
BY NC ND

Spatiotemporal Complex Covariance Functions for Vectorial Data

Sandra De Iaco

Abstract In the literature, complex-valued random fields were initially employed to characterize the spatial evolution of vector data, while the temporal dimension was typically treated separately or modeled using time-dependent complex covariance structures. Nevertheless, as in the real case, extending the complex formalism to a unified spatiotemporal framework and developing new classes of spatiotemporal complex covariance models are of clear interest to researchers, driven in part by the rapid increase in environmental vector observations collected over space and time. This contribution outlines the foundational elements of the complex spatiotemporal random field framework and describes how several spatiotemporal complex-valued covariance families can be derived from the spatial setting as well as from positive mixtures of infinite components or a convolution of the real part.

Keywords Vectorial data · Spatiotemporal covariance · Complex random fields · Complex covariance model

1 Introduction

The theoretical framework of the complex-valued random fields is reasonably considered appropriate for phenomena naturally decomposed into magnitude and direction, like wind, sea currents, and electric fields. In particular, [34] provided an early exposition for one-dimensional domains, while key spatial studies in this context include the contributions of [15, 18, 19, 32]. Some developments in parametric complex-domain covariance models appear in [26, 27] and some fitting

S. De Iaco (✉)
Department of Economic Sciences, University of Salento, Lecce, Italy

National Centre for HPC, Big Data and Quantum Computing, Bologna, Italy

National Biodiversity Future Center, Palermo, Italy
e-mail: sandra.deiaco@unisalento.it

S. De Iaco et al. (eds.), *Exploration of Spatio-Temporal Environmental Conditions: Harmonized Databases and Analytical Techniques*, Springer Proceedings in Mathematics & Statistics 531, https://doi.org/10.1007/978-3-032-17526-7_2

and computational issues can be found in [12]. Moreover the R package cgeostat [9] supports full analyses of 2D vector data.

The growing applied importance of complex space-time analysis drives demand for spatiotemporal tools. Many works treat spatiotemporal modeling for real scalar data [6, 8, 25, 29, 30, 33], but complex-domain extensions are scarce. Moreover, [2, 3] presented time-varying complex covariance functions whose parameters evolve over time, yet these do not model the joint space-time dependence, because they are not functions of temporal lags. In addition, [4] proposed a spatiotemporal complex random field characterization using Bochner's theorem. Indeed, the construction of wide families of spatiotemporal complex covariance models, for second-order stationary spatiotemporal complex-valued random fields, can be based on different techniques. The first two approaches, described in [11], propose an extension, to the space-time domain, of existing complex covariance models given by Lajaunie and Béjaoui [19], who built the imaginary component of a continuous complex covariance model on the basis of the real component through Radon-Nikodym's theorem and by De Iaco et al. [15], which started from the spectral representation of positive definite functions. Note that the latter includes by definition a periodic factor, whose effect can be damped through the product with a real monotonically decaying covariance model, while for the former the periodic component can be optionally incorporated in the model. The other approach allows for building a new family of complex covariance models through positive mixtures [10]. In particular, this family can be used to describe nonnegative real part or real and imaginary parts with or without a periodic component. This method generalizes the convolution-based class and enables flexible construction of spatiotemporal complex covariance families by defining real and imaginary parts using known space-time positive definite functions.

In this chapter, Sect. 2 introduces the concept of complex-valued space-time random fields, the second-order stationarity assumption, and the space-time complex covariance characterization. Section 3 provides an overview of spatiotemporal complex covariance families, with some examples, and discusses some advantages and further generalizations, while Sect. 4 presents some conclusions.

2 Theory of Spatio-temporal Complex Random Fields

A complex-valued random field in space-time is defined as follows:

$$Z(\mathbf{w}) = U(\mathbf{w}) + i\ V(\mathbf{w}), \qquad \mathbf{w} = (\mathbf{w}_s, w_t) \in \mathbb{R}^N \times \mathbb{R}, \tag{1}$$

where U and V are two real-valued spatiotemporal random fields, defined on $\mathbb{R}^N \times \mathbb{R}$, and $i = \sqrt{-1}$ is the imaginary number.

Assuming that Z has finite variance, the first-order moment $\mathbb{E}[Z(\mathbf{w})]$ of Z is finite and is defined through the following complex-valued deterministic function:

$$m(\mathbf{w}) = m_U(\mathbf{w}) + i\, m_V(\mathbf{w}),$$

where m_U and m_V are the expected values of U and V, respectively, whereas the covariance function $R(\mathbf{w}, \mathbf{w}')$ is finite and defined as

$$R(\mathbf{w}, \mathbf{w}') = \mathbb{E}\left[(Z(\mathbf{w}) - m(\mathbf{w}))(\overline{Z(\mathbf{w}') - m(\mathbf{w}')})\right] = \quad (2)$$

$$= R_U(\mathbf{w}, \mathbf{w}') + R_V(\mathbf{w}, \mathbf{w}') + i\, [R_{VU}(\mathbf{w}, \mathbf{w}') - R_{UV}(\mathbf{w}, \mathbf{w}')], \qquad \mathbf{w}, \mathbf{w}' \in \mathbb{R}^N \times \mathbb{R},$$

where R_U and R_V denote, respectively, the spatiotemporal covariance functions associated with the real part U and the imaginary part V of Z, whereas R_{VU} and R_{UV} represent the related cross-covariances. As usually done, the complex conjugation is denoted with an overline in Eq. (2).

The complex formalism differs conceptually from multivariate Geostatistics; indeed, it is recalled when data admit a complex representation. Vector measurements like wind or ocean currents are naturally handled as complex-valued fields. The key modeling benefit is that the complex covariance fully characterizes the two-dimensional correlation, where the real part captures direct covariances and the imaginary part captures cross-covariances, so only the sum of direct covariances and the difference of cross-covariances need to be modeled. This advantage grows for spatiotemporal data. Further details on this aspect can be found in [12, 32].

2.1 Second-Order Stationarity Hypothesis

Given a random field as in (1), it is second-order stationary in space-time, if its complex-valued expected value is constant, that is, $\mathbb{E}[Z(\mathbf{w})] = m = m_U + i\, m_V$, where the expected values of U and V are constant for any point of the domain and the space-time covariance function R depends, for any pair of points $\mathbf{w} = (\mathbf{w}_s, w_t)$ and $\mathbf{w}' = (\mathbf{w}'_s, w'_t)$, on the lag separation vector $\mathbf{h}_s = \mathbf{w}'_s - \mathbf{w}_s \in \mathbb{R}^N$, $h_t = w_t - w'_t \in \mathbb{R}$, that is,

$$R(\mathbf{h}_s, h_t) = R^{re}(\mathbf{h}_s, h_t) + i\, R^{im}(\mathbf{h}_s, h_t), \quad (3)$$

where the corresponding real part, indicated with R^{re}, is given as follows:

$$R^{re}(\mathbf{h}_s, h_t) = R_U(\mathbf{h}_s, h_t) + R_V(\mathbf{h}_s, h_t), \quad (4)$$

and the imaginary part, denoted with R^{im}, is defined as

$$R^{im}(\mathbf{h}_s, h_t) = R_{VU}(\mathbf{h}_s, h_t) - R_{UV}(\mathbf{h}_s, h_t). \quad (5)$$

Note that, by definition, the even function in Eq. (4) is also a space-time covariance function on the basis of the convexity property [5]; on the other hand, the odd function in Eq. (5) is not a space-time covariance function.

For a second-order stationary complex-valued random field Z, the space-time covariance function R satisfies the following properties, i.e.:

- It is nonnegative at zero, $R(\mathbf{0}, 0) \geq 0$.
- It is symmetric, $R(-\mathbf{h}_s, -h_t) = \overline{R(\mathbf{h}_s, h_t)}$, that is, the covariance matrix is Hermitian.
- It is bounded, $|R(\mathbf{h}_s, h_t)| \leq R(\mathbf{0}, 0)$.

Moreover, any complex covariance function must be nonnegative definite, i.e.,

$$\sum_{i=1}^{k} \sum_{j=1}^{k} R(\mathbf{w}_i - \mathbf{w}_j) a_i \overline{a_j} \geq 0, \tag{6}$$

$\forall\, \mathbf{w}_1, \mathbf{w}_2, \ldots, \mathbf{w}_k \in \mathbb{R}^N \times \mathbb{R}, \quad \forall\, a_1, \ldots, a_k \in \mathbb{C}$.

2.2 Bochner's Theorem and Space-Time Complex Covariance

On the basis of the full characterization given for any continuous complex covariance function by Bochner's theorem [1], the definition of a space-time complex covariance function is provided as follows:

$$R(\mathbf{h}_s, h_t) = \int_{\mathbb{R}^N \times \mathbb{R}} \cos(\mathbf{h}_s \cdot \mathbf{q}_s + h_t q_t) dF(\mathbf{q}_s, q_t) + i \int_{\mathbb{R}^N \times \mathbb{R}} \sin(\mathbf{h}_s \cdot \mathbf{q}_s + h_t q_t) dF(\mathbf{q}_s, q_t), \tag{7}$$

where $\mathbf{h}_s \cdot \mathbf{q}_s$ denotes the inner product $\sum_{i=1}^{N} h_i q_i$ and F, called the *space-time spectral distribution function* of R, is a bounded, real-valued function and satisfies the following condition:

$$\int_A dF(\mathbf{q}_s, q_t) \geq 0, \qquad \text{for any measurable } A \subseteq \mathbb{R}^N \times \mathbb{R}.$$

Thus, the space-time covariance function of Z can be written as the sum of an even part and an odd one. If the spectral distribution function F is absolutely continuous, the space-time spectral density f exists and is a bounded and continuous nonnegative function. In particular, if the spectral distribution F is symmetric about the origin of $\mathbb{R}^N \times \mathbb{R}$ or if a spectral density function f is an even function, then the covariance function is real-valued. If the spectral density is separable, i.e., $f(\mathbf{q}_s, q_t) = f_1(\mathbf{q}_s) f_2(q_t)$, then the space-time complex covariance function is separable:

$$R(\mathbf{h}_s, h_t) = \int_{\mathbb{R}^N} e^{i(\mathbf{h}_s \cdot \mathbf{q}_s)} f_1(\mathbf{q}_s) d\mathbf{q}_s \int_{\mathbb{R}} e^{i h_t q_t} f_2(q_t) dq_t = R_1(\mathbf{h}_s) R_2(h_t), \tag{8}$$

where R_1 and R_2 are spatial and temporal complex covariance functions, respectively. Thus, two different subclasses can be derived by assuming that one of the spectral densities, f_1 or f_2, is symmetric. Note that, only for the real case, the covariance function R is an even function, since the imaginary part of (3) is zero.

3 Some Complex Families from Space to Space-Time

Spatiotemporal complex covariance families on $\mathbb{R}^n \times \mathbb{R}$ can be derived by extending spatial models on $\mathbb{R}^N$. In the following, some classes, obtained by using the Radon-Nikodym theorem [19] or on the basis of spectral-density translation [15] as well as through positive mixtures, have been presented, as discussed in [4, 11].

3.1 Convolution of the Real Part

One of the earliest contributions to complex covariance models on spatial domains was by Lajaunie and Béjaoui [19], who constructed the imaginary part directly from the real part. Given the real component $R^{re}(\mathbf{h}_s, h_t)$, the imaginary part follows from the Radon-Nikodym theorem, as explained below:

$$i R^{im}(\mathbf{h}_s, h_t) = (F * R^{re})(\mathbf{h}_s, h_t), \tag{9}$$

where $*$ represents the convolution operation and F is a complex distribution whose Fourier transform is a real odd function f with values in the interval $[-1, 1]$. Thus, the following class of complex spatiotemporal covariance model can be obtained

$$R(\mathbf{h}_s, h_t) = R^{re}(\mathbf{h}_s, h_t) + (F * R^{re})(\mathbf{h}_s, h_t). \tag{10}$$

The imaginary components $R^{im}(\mathbf{h}_s, h_t)$, linked to a real continuous covariances $R^{re}(\mathbf{h}_s, h_t)$ by Eq. (10), are called compatible imaginary parts. A parametric family of such compatible imaginary components can be constructed as follows:

$$R^{im}(\boldsymbol{h}_s, h_t) = 0.5 \int [R^{re}(\boldsymbol{h}_s - \boldsymbol{\tau}_s, h_t - \tau_t) - R^{re}(\boldsymbol{h}_s + \boldsymbol{\tau}_s, h_t + \tau_t)] \mu(d\boldsymbol{\tau}_s, d\tau_t), \tag{11}$$

where μ is a real bounded measure such that $|\mu(\omega_s, \omega_t)| < 1$ with $f(\boldsymbol{\omega}_s, \omega_t) = -\int \sin(\boldsymbol{\omega}_s \cdot \boldsymbol{\tau}_s, \omega_t \tau_t) \mu(d\boldsymbol{\tau}_s, d\tau_t)$. In other terms, the imaginary parts of this class can be got through the expected value of the difference $[R^{re}(\mathbf{h}_s - \boldsymbol{\tau}_s, h_t - \tau_t) - R^{re}(\mathbf{h}_s + \boldsymbol{\tau}_s, h_t + \tau_t)\mu(d\boldsymbol{\tau}_s, d\tau_t)]$ with respect to $\boldsymbol{\tau}_s$ and τ_t, i.e.,

$$R^{im}(\mathbf{h}_s, h_t) = 0.5\mathbb{E}_{\boldsymbol{\tau}_s,\tau_t}[R^{re}(\mathbf{h}_s - \boldsymbol{\tau}_s, h_t - \tau_t) - R^{re}(\mathbf{h}_s + \boldsymbol{\tau}_s, h_t + \tau_t)]. \qquad (12)$$

As a special case of (12), it is worth introducing the following simple complex covariance model in space-time:

$$R(\mathbf{h}_s, h_t) = R^{re}(\mathbf{h}_s, h_t) + 0.5i[R^{re}(\mathbf{h}_s - \boldsymbol{\tau}_s, h_t - \tau_t) - R^{re}(\mathbf{h}_s + \boldsymbol{\tau}_s, h_t + \tau_t)], \qquad (13)$$

where the imaginary component is determined through the difference of the real-valued covariance function R^{re} translated backward and forward of the vector $(\boldsymbol{\tau}_s, \tau_t)$. An example of a class of compatible imaginary parts associated with the Gneiting class of real-valued covariance functions has been proposed below.

Example 1 Suppose that the real component of Eq. (13) is described by the following class of Gneiting covariance models:

$$R^{re}_{Gn}(\mathbf{h}_s, h_t) = \frac{1}{(a|h_t|^{2\alpha} + 1)} \exp - \frac{b\|\mathbf{h}_s\|^{2\gamma}}{(a|h_t|^{2\alpha} + 1)^{\gamma\beta}}, \qquad (14)$$

with $a, b \in]0, +\infty]$, $\alpha, \gamma, \beta \in]0, 1]$; then the corresponding compatible imaginary part can be obtained on the basis of the abovementioned difference, that is,

$$R^{im}_{Gn}(\mathbf{h}_s, h_t) = \frac{0.5}{(a|h_t - \tau_t|^{2\alpha} + 1)} \exp \frac{-b\|\mathbf{h}_s - \boldsymbol{\tau}_s\|^{2\gamma}}{(a|h_t - \tau_t|^{2\alpha} + 1)^{\gamma\beta}} + \\ - \frac{0.5}{(a|h_t + \tau_t|^{2\alpha} + 1)} \exp \frac{-b\|\mathbf{h}_s + \boldsymbol{\tau}_s\|^{2\gamma}}{(a|h_t + \tau_t|^{2\alpha} + 1)^{\gamma\beta}}. \qquad (15)$$

Figure 1 provides a graphical representation of the spatial and temporal marginals of the complex model obtained from Eqs. (14) and (15), that is, the real marginals in space and time

$$R^{re}_{Gn}(\mathbf{h}_s, 0) = \frac{1}{(a|h_t|^{2\alpha} + 1)}, \qquad R^{re}_{Gn}(\mathbf{0}, h_t) = \exp[-b\|\mathbf{h}_s\|^{2\gamma}], \qquad (16)$$

and the imaginary marginals in space and time

$$R^{im}_{Gn}(\mathbf{h}_s, 0) = 0.5/a' \left\{ \exp[-b/a'^{\gamma\beta}\|\mathbf{h}_s - \boldsymbol{\tau}_s\|^{2\gamma}] - \exp[-b/a'^{\gamma\beta}\|\mathbf{h}_s + \boldsymbol{\tau}_s\|^{2\gamma}] \right\}, \qquad (17)$$

$$R^{im}_{Gn}(\mathbf{0}, h_t) = \frac{0.5}{(a|h_t - \tau_t|^{2\alpha} + 1)} \exp[-b'/(a|h_t - \tau_t|^{2\alpha} + 1)^{\gamma\beta}] +$$

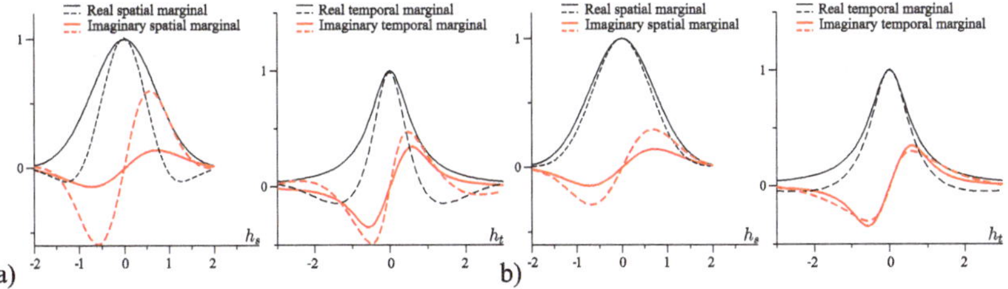

Fig. 1 Real and imaginary marginals of the complex models based on: the construction given in (13), written in terms of the Gneiting class in Eq. (14), with $\alpha = \gamma = 1, a = 2.5, b = 1, \beta = 1$, and $\boldsymbol{\tau}_s = (0.2, 0.2)$, $\tau_t = 0.5$ (continuous line) and the construction in (24), with $\tilde{R}$ defined as in Eq. (14) with the same parameter values $\alpha = \gamma = 1, a = 2.5, b = 1, \beta = 1$ (dashed line), and (**a**) $\mathbf{c}_s = (1.5, 1.5), c_t = 1.8$, (**b**) $\mathbf{c}_s = (0.5, 0.5), c_t = 1$

$$- \frac{0.5}{(a|h_t + \tau_t|^{2\alpha} + 1)} \exp[-b'/(a|h_t + \tau_t|^{2\alpha} + 1)^{\gamma\beta}], \tag{18}$$

where $a' = (a|\tau_t|^{2\alpha} + 1)$, $b' = b\|\boldsymbol{\tau}_s\|^{2\gamma}$.

It is clear that any periodic component can be included, for example, through the sum or the product of the class in Eq. (13) with the basic complex covariance function, that is, $R(\mathbf{h}_s, h_t) = \cos(\mathbf{h}_s \cdot \mathbf{c}_s + h_t c_t) + i\ \sin(\mathbf{h}_s \cdot \mathbf{c}_s + h_t c_t)$, or just with the real part of the same basic complex covariance function.

3.2 Construction Based on Asymmetric Spectral Density

Starting from Bochner's characterization, a first attempt to construct complex covariance models for the spatial case was proposed by De Iaco et al. [15], through a translation of an even spectral density function. In [4], the above approach has been suitably generalized to build a class of complex-valued spatiotemporal covariance models. Indeed, given a spectral representation of a real-valued covariance function $\tilde{R}(\mathbf{h}_s, h_t)$ and the corresponding even spectral density function $f(\mathbf{q}_s, q_t)$, then the translated function $f(\mathbf{q}_s - \mathbf{c}_s, q_t - c_t)$, which is a not-even spectral density function $\forall\, \mathbf{c} = (\mathbf{c}_s, c_t) \in \mathbb{R}^N \times \mathbb{R}$, $\mathbf{c} = (\mathbf{c}_s, c_t) \neq \mathbf{0}$, is used to define the following class of space-time complex covariance models:

$$\begin{aligned} R(\mathbf{h}_s, h_t; \mathbf{c}) &= \int_{\mathbb{R}^N \times \mathbb{R}} \exp[i(\mathbf{h}_s \cdot \mathbf{q}_s + h_t q_t)] f(\mathbf{q}_s - \mathbf{c}_s, q_t - c_t) d\mathbf{q}_s dq_t = \\ &= \exp[i(\mathbf{h}_s \cdot \mathbf{c}_s + h_t c_t)]\tilde{R}(\mathbf{h}_s, h_t). \end{aligned} \tag{19}$$

For modeling purposes, the above class can be better made explicit as follows:

$$R(\mathbf{h}_s, h_t; \mathbf{c}) = \cos(\mathbf{h}_s \cdot \mathbf{c}_s + h_t c_t)\tilde{R}(\mathbf{h}_s, h_t) + i\ \sin(\mathbf{h}_s \cdot \mathbf{c}_s + h_t c_t)\tilde{R}(\mathbf{h}_s, h_t), \tag{20}$$

where $\tilde{R}(\mathbf{h}_s, h_t)$ is a real space-time covariance function and $\mathbf{c} = (\mathbf{c}_s, c_t) \in \mathbb{R}^N \times \mathbb{R}$ represents the shifting factor.

Looking simultaneously to Eqs. (3) and (20), it is worth pointing out that the real part $\cos(\mathbf{h}_s \cdot \mathbf{c}_s + h_t c_t)\tilde{R}(\mathbf{h}_s, h_t)$ describes the sum of the covariances $R_U(\mathbf{h}_s, h_t)$ and $R_V(\mathbf{h}_s, h_t)$, while the imaginary part $\sin(\mathbf{h}_s \cdot \mathbf{c}_s + h_t c_t)\tilde{R}(\mathbf{h}_s, h_t)$ is a model for the difference of the two cross-covariances R_{VU} and R_{UV}.

Note that the spatial and temporal complex marginal covariance models are

$$R(\mathbf{h}_s, 0; \mathbf{c}) = \cos(\mathbf{h}_s \cdot \mathbf{c}_s)\tilde{R}(\mathbf{h}_s, 0) + i\ \sin(\mathbf{h}_s \cdot \mathbf{c}_s)\tilde{R}(\mathbf{h}_s, 0), \tag{21}$$

$$R(\mathbf{0}, h_t; \mathbf{c}) = \cos(h_t c_t)\tilde{R}(\mathbf{0}, h_t) + i\ \sin(h_t c_t)\tilde{R}(\mathbf{0}, h_t), \tag{22}$$

which can be very useful under a computational point of view, by following the same idea given in [9, 12]. A discussion on the effect of the shifting factor can be found in [16].

Note that it is common to adopt real-valued spatiotemporal covariance models which are isotropic only in space; thus the following space-time complex covariance can be introduced as a special case of (20):

$$R(\mathbf{h}_s, h_t; \mathbf{c}) = \cos(\mathbf{h}_s \cdot \mathbf{c}_s + h_t c_t)\tilde{R}(\|\mathbf{h}_s\|, h_t) + i\ \sin(\mathbf{h}_s \cdot \mathbf{c}_s + h_t c_t)\tilde{R}(\|\mathbf{h}_s\|, h_t). \tag{23}$$

In the following examples, some spatiotemporal models on $\mathbb{R}^N \times \mathbb{R}$, with $N = 2$ and vector coordinate $(\mathbf{h}_s, h_t) = (h_x, h_y, h_t)^T$, are presented.

Example 2 Let $\tilde{R}$ be the class of Gneiting covariance models in Eq. (14); then given the complex class in (23), the following space-time complex covariance model can be built:

$$\begin{aligned} R(\mathbf{h}_s, h_t) = &\frac{\cos(\mathbf{h}_s \cdot \mathbf{c}_s + h_t c_t)}{(a|h_t|^{2\alpha} + 1)} \exp\left[-\frac{b\|\mathbf{h}_s\|^{2\gamma}}{(a|h_t|^{2\alpha} + 1)^{\gamma\beta}}\right] + \\ &+ i\ \frac{\sin(\mathbf{h}_s \cdot \mathbf{c}_s + h_t c_t)}{(a|h_t|^{2\alpha} + 1)} \exp\left[-\frac{b\|\mathbf{h}_s\|^{2\gamma}}{(a|h_t|^{2\alpha} + 1)^{\gamma\beta}}\right]. \end{aligned} \tag{24}$$

Note that the real-valued covariance component $\tilde{R}(\mathbf{h}_s, h_t)$ is assumed to be isotropic in space. Moreover, the spatial and temporal complex marginal covariance models are defined as follows:

$$R(\mathbf{h}_s, 0) = \cos(\mathbf{h}_s \cdot \mathbf{c}_s)\exp[-b\|\mathbf{h}_s\|^{2\gamma}] + i\ \sin(\mathbf{h}_s \cdot \mathbf{c}_s)\exp[-b\|\mathbf{h}_s\|^{2\gamma}], \tag{25}$$

$$R(\mathbf{0}, h_t) = \frac{\cos(h_t c_t)}{(a|h_t|^{2\alpha} + 1)} + i\ \frac{\sin(h_t c_t)}{(a|h_t|^{2\alpha} + 1)}. \tag{26}$$

Figure 1 provides a graphical representation of the real and imaginary parts of the spatial and temporal marginals of the complex model in (24), together with the marginals of the model in Eq. (13) for comparison purposes. Indeed, both models are based on the Gneiting class defined in Eq. (14), with the same values for the parameters. It is clear the effect of the factors containing the shifting vector $\mathbf{c}$; however, the $\tilde{R}$ produces a damping effect for both the real and imaginary components of the spatial and temporal marginals. The damped periodicity might be further mitigated by using lower values for the shifting factor, as in Fig. 1b.

3.3 Infinite Positive Power Mixture

As will be clarified, other families of complex-valued spatiotemporal covariance models can be generated through a combination of an infinite number of terms, as in (20), weighted with a power of a constant in the interval $]0, 1[$ [10].

Theorem 1 *Let $\tilde{R}(\mathbf{h}_s, h_t)$ be a real-valued covariance function and $a \in]0, 1[$; then*

$$R(\mathbf{h}_s, h_t; a, \mathbf{c}) = \frac{1 - a\cos(\mathbf{h}_s \cdot \mathbf{c}_s + h_t c_t)}{1 - 2a\cos(\mathbf{h}_s \cdot \mathbf{c}_s + h_t c_t) + a^2}\tilde{R}(\mathbf{h}_s, h_t) + i\,\frac{a\sin(\mathbf{h}_s \cdot \mathbf{c}_s + h_t c_t)}{1 - 2a\cos(\mathbf{h}_s \cdot \mathbf{c}_s + h_t c_t) + a^2}\tilde{R}(\mathbf{h}_s, h_t) \tag{27}$$

is a class of complex-valued covariance models.

One of the advantages of this class is related to the minimum and the maximum of the factors

$$\frac{1 - a\cos(\mathbf{h}_s \cdot \mathbf{c}_s + h_t c_t)}{1 - 2a\cos(\mathbf{h}_s \cdot \mathbf{c}_s + h_t c_t) + a^2}, \tag{28}$$

$$\frac{a\sin(\mathbf{h}_s \cdot \mathbf{c}_s + h_t c_t)}{1 - 2a\cos(\mathbf{h}_s \cdot \mathbf{c}_s + h_t c_t) + a^2}, \tag{29}$$

which are different for the real and the imaginary parts, that is, $[1/(1+a), 1/(1-a)]$ for (28) and $[-a/(1-a^2), a/(1-a^2)]$ for (29).

Moreover, apart from the amplitude of the periodic component of this class, which can be smoothed by fixing low values of the parameter a, it is worth pointing out the flexibility of the class (27), with respect to (20), in describing real and imaginary parts which are not affected by the presence of periodic components, as shown in the following example.

Example 3 Let $\tilde{R}$ be described by the integrated class, obtained by integrating the stable model with the gamma function, that is,

$$\tilde{R}(\mathbf{h}_s, h_t) = \frac{1}{(b_t|h_t|^{2\alpha} + b_s\|\mathbf{h}_s\|^{2\gamma} + 1)}, \tag{30}$$

with $b_s, b_t \in]0, +\infty]$ and $\alpha = \gamma \in]0, 1]$; then from (27), the obtained complex spatiotemporal covariance function is given below:

$$R(\mathbf{h}_s, h_t; \boldsymbol{\theta}) = R^{re}(\mathbf{h}_s, h_t; \boldsymbol{\theta}) + i\, R^{im}(\mathbf{h}_s, h_t; \boldsymbol{\theta}), \tag{31}$$

where the real and imaginary parts are, respectively,

$$R^{re}(\mathbf{h}_s, h_t; \boldsymbol{\theta}) = \frac{1 - a\cos(\mathbf{h}_s \cdot \mathbf{c}_s + h_t c_t)}{(1 - 2a\cos(\mathbf{h}_s \cdot \mathbf{c}_s + h_t c_t) + a^2)(b_t |h_t|^{2\alpha} + b_s \|\mathbf{h}_s\|^{2\gamma} + 1)}, \tag{32}$$

$$R^{im}(\mathbf{h}_s, h_t; \boldsymbol{\theta}) = \frac{a\sin(\mathbf{h}_s \cdot \mathbf{c}_s + h_t c_t)}{(1 - 2a\cos(\mathbf{h}_s \cdot \mathbf{c}_s + h_t c_t) + a^2)(b_t |h_t|^{2\alpha} + b_s \|\mathbf{h}_s\|^{2\gamma} + 1)}, \tag{33}$$

with $\boldsymbol{\theta} = (a, b_s, b_t, \mathbf{c}, \alpha, \gamma)$. Figure 2a, b provides a graphical comparison between the spatial and temporal marginals of the classes (20) and (31), where the parameters related to the shifting factors in space and time are assumed to be the same for the two classes. The amplitude of the periodic component of the class (31) is smoothed by fixing a value of the parameter a less than 0.5. Moreover, Fig. 2c, d illustrates the flexibility of the class (31), with respect to (20), in describing real and imaginary parts which are not characterized by periodic components. It is clear that the class (31) is able to describe random fields whose correlation decays monotonously, while

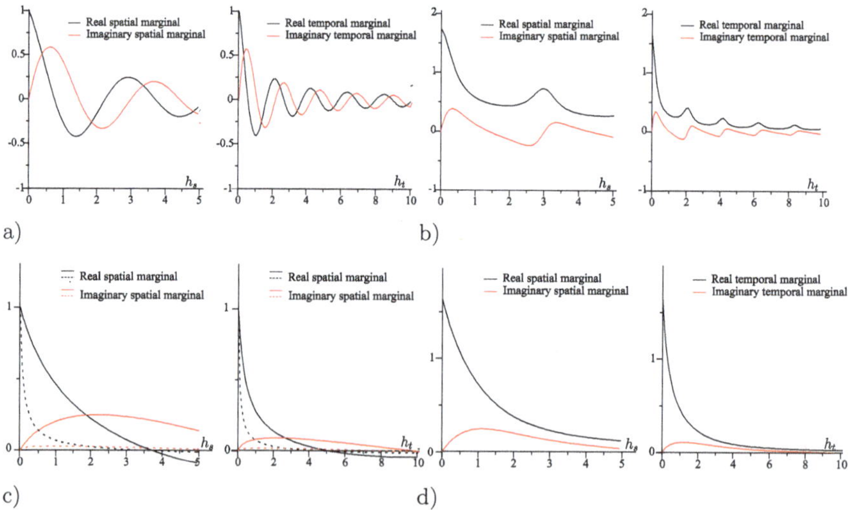

Fig. 2 Real and imaginary parts of the spatial and temporal marginals of the complex models, based on the construction given in **(a)** (20) with $(\mathbf{c}_s, c_t) = (1.5, 1.5, 3)$, **(b)** (31), with $(\mathbf{c}_s, c_t) = (1.5, 1.5, 3)$ and $a = 0.4$, where $\tilde{R}$ is defined in Eq. (30), with $(b_s, b_t, \alpha, \gamma) = (1, 0.5, 0.5, 0.5)$, **(c)** (20), with $(\mathbf{c}_s, c_t) = (0.3, 0.3, 0.3)$ and $\tilde{R}$ as in Eq. (30) with $(b_s, b_t, \alpha, \gamma) = (1, 2.5, 0.5, 0.5)$ (continuous line) or $\tilde{R}$ as in Eq. (30) with $(b_s, b_t, \alpha, \gamma) = (12, 12, 0.5, 0.5)$ (dashed line), and **(d)** (31), with $(\mathbf{c}_s, c_t) = (0.3, 0.3, 0.3)$, $a = 0.4$, and $(b_s, b_t, \alpha, \gamma) = (1, 2.5, 0.5, 0.5)$

for the class (20) the effect of the periodic component can be left out by keeping low the $\tilde{R}$ parameters associated with the ranges, as shown by the dashed lines in Fig. 2a.

3.4 Advantages and Further Generalizations

The potentiality of the proposed approaches is related to the flexibility of building wide families of spatiotemporal complex-valued covariance functions, by easily defining the real and imaginary parts exploiting what is already known on real-valued positive definite functions in space-time. Indeed, the real component used in the convolution-based class, given in Eq. (13), or the real-valued covariance factor in the construction based on asymmetric spectral density, as in (20), as well as the real-valued covariance factor in the construction through a positive mixture, as in (27), can be explicated by selecting a model within a wide range of separable and non-separable spatiotemporal classes of covariance models available in the literature [7, 13, 14, 17, 20–24, 28, 29, 31]. Moreover, the classes proposed in this section can be further used to generalize the convolution-based class in Eq. (13), as specified below.

Let $R^{re}(\mathbf{h}_s, h_t)$ be a real-valued covariance function and $a \in]0, 1[$; then

$$R(\mathbf{h}_s, h_t; a, \mathbf{c}) = \frac{1 - a\cos(\mathbf{h}_s \cdot \mathbf{c}_s + h_t c_t)}{1 - 2a\cos(\mathbf{h}_s \cdot \mathbf{c}_s + h_t c_t) + a^2} R^{re}(\mathbf{h}_s, h_t) +$$

$$+ 0.5i \frac{1 - a\cos(\mathbf{h}_s \cdot \mathbf{c}_s + h_t c_t)}{1 - 2a\cos(\mathbf{h}_s \cdot \mathbf{c}_s + h_t c_t) + a^2}$$

$$[R^{re}(\mathbf{h}_s - \boldsymbol{\tau}_s, h_t - \tau_t) - R^{re}(\mathbf{h}_s + \boldsymbol{\tau}_s, h_t + \tau_t)] \tag{34}$$

is a class of complex valued covariance models.

The convolution-based class and its generalization, the class based on asymmetric spectral density and the class obtained through a positive mixture cover an extensive range of complex correlation structures, whose real part (not necessarily nonnegative) and imaginary part do/do not exhibit a periodic component. In particular, the convolution-based class can be easily applied to describe complex covariance surfaces with a nonnegative real part, the class based on asymmetric spectral density is marked by a periodic factor for both the real and imaginary components, whose effect is often damped through the product with a real monotonically decaying covariance model. Thus, they can be typically used for modeling complex covariance surfaces, whose real part can assume negative values. Differently from the other two classes, the family of the mixture models presents a real part which is nonnegative unless the model used for the real-valued covariance factor assumes negative values; moreover, the amplitude and the frequency of the periodic factor can be conveniently fixed according to the empirical behavior of the real and imaginary components, or if necessary, the effect of the periodic factor can be left out, without producing a contextual flattening of the imaginary part as happens for the class based on translation of a symmetric spectral density. Some details on the estimation and modeling aspects and some applications on oceanographic data can be found in [10, 11].

4 Conclusions

In this contribution, the formalism of complex-valued random fields was introduced in a space-time domain and some classes of spatiotemporal complex covariance models were presented. However, these advances of the complex-valued random fields in space-time can be considered a good starting point for future works, focused on further theoretical developments and practical aspects which can also unleash the potential in environmental applications.

Bibliography

1. Bochner, S.: Monotone funktionen, stieltjessche integrale und harmonische analyse. Math. Ann. **108**(1), 378–410 (1933)
2. Cappello, C., De Iaco, S., Maggio, S., Posa, D.: Modeling ocean currents through complex random fields indexed in time. Math. Geosci. **53**, 999–1025 (2020)
3. Cappello, C., De Iaco, S., Maggio, S., Posa, D.: Time varying complex covariance functions for oceanographic data. Spatial Stat. **42**(4), 100426 (2020)
4. Cappello, C., De Iaco, S., Maggio, S., Posa, D.: Modeling spatio-temporal complex covariance functions for vectorial data. Spatial Stat. **47**, 100562 (2022)
5. Chilès, J.P., Delfiner, P.: Geostatistics: Modeling Spatial Uncertainty, 2 edn. J. Wiley & Sons, New York (2012)
6. Christakos, G.: Spatio-temporal Random Fields: Theory and Applications. Elsevier Science Publishing Co Inc., Amsterdam (2017)
7. Cressie, N., Huang, H.: Classes of nonseparable, spatio-temporal stationary covariance functions. J. Am. Stat. Assoc. **94**(448), 1330–1340 (1999)
8. Cressie, N., Wikle, C.K.: Statistics for Spatio-Temporal Data. J. Wiley & Sons Inc., Hoboken, New Jersey (2011)
9. De Iaco, S.: The cgeostat software for analyzing complex-valued random fields. J. Stat. Softw. **79**(5), 1–32 (2017)
10. De Iaco, S.: New spatio-temporal complex covariance functions for vectorial data through positive mixtures. Stochastic Environ. Res. Risk Assess. **36**, 2769–2787 (2022)
11. De Iaco, S.: Spatio-temporal generalized complex covariance models based on convolution. Comput. Stat. Data Anal. **183**, 107709 (2023)
12. De Iaco, S., Posa, D.: Wind velocity prediction through complex kriging: formalism and computational aspects. Environ. Ecol. Stat. **23**(1), 115–139 (2016)
13. De Iaco, S., Myers, D.E., Posa, D.: Space-time analysis using a general product-sum model. Stat. Probab. Lett. **52**(1), 21–28 (2001)
14. De Iaco, S., Myers, D.E., Posa, D.: Nonseparable space-time covariance models: some parametric families. Math. Geol. **34**(1), 23–42 (2002)
15. De Iaco, S., Palma, M., Posa, D.: Covariance functions and models for complex-valued random fields. Stochastic Environ. Res. Risk Assess. **17**(3), 145–156 (2003)
16. De Iaco, S., Posa, D., Palma, M.: Complex-valued random fields for vectorial data: estimating and modeling aspects. Math. Geosci. **45**(5), 557–573 (2013)

17. Gneiting, T.: Nonseparable, stationary covariance functions for space-time data. J. Am. Stat. Assoc. **97**(458), 590–600 (2002)
18. Grzebyk, M.: Ajustement d'une Coregionalisation Stationnaire. Ph.D. thesis, Ecoles des Mines, Paris (1993)
19. Lajaunie, C., Béjaoui, R.: Sur le krigeage des fonctions complexes. Tech. Rep. N-23/91/G, Centre de Geostatistique, Ecole des Mines de Paris, Fontainebleau (1991)
20. Ma, C.: Spatio-temporal covariance functions generated by mixtures. Math. Geol. **34**, 965–975 (2002)
21. Ma, C.: Families of spatio-temporal stationary covariance models. J. Stat. Plan. Infer. **116**, 489–501 (2003)
22. Ma, C.: Linear combinations of space-time covariance functions and variograms. IEEE Trans. Signal Proces. **53**(3), 857–864 (2005)
23. Porcu, E., Gregori, P., Mateu, J.: Nonseparable stationary anisotropic spacetime covariance functions. Stochastic Environ. Res. Risk Assess. **21**(2), 113–122 (2006)
24. Porcu, E., Mateu, J., Saura, F.: New classes of covariance and spectral density functions for spatio-temporal modelling. Stochastic Environ. Res. Risk Assess. **22**(Suppl 1), 65–79 (2008)
25. Posa, D.: A simple description of spatio-temporal processes. Comput. Stat. Data Anal. **15**(4), 425–437 (1993)
26. Posa, D.: Parametric families for complex valued covariance functions: some results, an overview and critical aspects. Spatial Stat. **39**, 100473 (2020)
27. Posa, D.: Models for the difference of continuous covariance functions. Stochastic Environ. Res. Risk Assess. **35**, 1369–1386 (2021)
28. Rodrigues, A., Diggle, P.J.: A class of convolution-based models for spatio-temporal processes with non-separable covariance structure. Scand. J. Stat. **37**(4), 553–567 (2010)
29. Rouhani, S., Hall, T.J.: Space-time kriging of groundwater data. In: Armstrong, M. (ed.) Geostatistics. Quantitative Geology and Geostatistics, vol. 4, pp. 639–650. Springer, Dordrecht (1989)
30. Stein, M.L.: A simple model for spatial-temporal processes. Water Resour. Res. **22**(13), 2107–2110 (1986)
31. Stein, M.L.: Space-time covariance functions. J. Am. Stat. Assoc. **100**(469), 310–321 (2005)
32. Wackernagel, H.: Multivariate geostatistics: an introduction with applications. Springer Series in Statistics. Springer, Berlin (2003)
33. Wikle, C.K., Zammit-Mangion, A., Cressie, N.: Spatio-Temporal Statistics with R. Chapman & Hall/CRC, Boca Raton (2019)
34. Yaglom, A.M.: Correlation theory of stationary and related random functions. Springer Series in Statistics, vols. I, II. Springer, Berlin (1987)

Non-separable Covariance Kernels for Spatiotemporal Gaussian Processes Based on the Hybrid Spectral Method

Dionissios T. Hristopulos

Abstract In the kriging and Gaussian process regression frameworks, the covariance kernel plays a fundamental role. The construction of flexible spatiotemporal regression kernels is essential for capturing dynamic physical processes. Hence, kernels derivable from stochastic equations of motion are in high demand for their effectiveness in modeling complex systems. We review the few cases of known purely spatial covariance kernels derived from stochastic differential equations. On the other hand, constructing non-separable spatiotemporal kernels derivable from physical equations presents significant challenges. We focus on the recently developed hybrid spectral method which combines temporal dependence derived from physical systems with observation-informed dispersion relations.

Keywords Spatial statistics · Gaussian process regression · Spatiotemporal kernels · Differential equations · Ornstein-Uhlenbeck

1 Introduction

Separable covariance models are often used in spatiotemporal learning: they are simple, computationally efficient, and help to simplify parameter estimation [5]. On the other hand, separability is not supported by solutions of partial differential equations that include stochastic terms [13]. In addition, separable models imply significantly simplified correlation structure for the underlying stochastic processes [18].

Non-separable covariance kernels can capture space-time interactions and provide improved predictive performance, physical interpretation, and flexibility in modeling complex correlation structures [3, 14, 17]. However, obtaining explicit expressions for spatiotemporal kernels that are derivable from physical equations of

D. T. Hristopulos (✉)
School of Electrical and Computer Engineering, Technical University of Crete, Chania, Greece
e-mail: dchristopoulos@tuc.gr

S. De Iaco et al. (eds.), *Exploration of Spatio-Temporal Environmental Conditions: Harmonized Databases and Analytical Techniques*, Springer Proceedings in Mathematics & Statistics 531, https://doi.org/10.1007/978-3-032-17526-7_3

motion is not trivial. Such solutions require solving partial differential equations which, even in the case of uncomplicated boundary and initial conditions, is a challenging task.

One data-driven approach is proper orthogonal decomposition (POD), a method used in the analysis of fluid flow, in which space-time covariance kernels are modeled as superposition of orthonormal functions that capture both spatial and temporal variations, similar to the Karhunen-Loève expansion [19, 20]. Each of the modes could represent separable space and time components, but the superposition is a non-separable function of space and time.

This chapter aims to briefly review the current state of development of kernels derivable from differential equations and then focus on the hybrid spectral approach. The remainder of this chapter has the following structure: Sect. 2 reviews the generation of purely spatial kernels. Section 3 discusses the challenges related to deriving space-time covariance functions linked to equations of motion (such as stochastic partial differential equations). Section 4 specializes in the hybrid spectral approach (HSA). Finally, Sect. 5 presents our conclusions.

2 Spatial Covariance Kernels from SPDEs

In purely spatial problems, explicit solutions for covariance kernels are available for three stochastic partial differential equations (SPDEs): the De Wijs model, the Whittle-Matérn SPDE [15], and SPDEs that involve polynomials of the half-Laplacian operator [9]. In the following, $\mathbf{s} \in \mathcal{D} \subset \mathbb{R}^d$ denotes the spatial location vector in a d-dimensional domain, while $\mathbf{k}$ denotes the d-dimensional wavevector (spatial frequency vector) in the reciprocal (Fourier) space. We use $X(\mathbf{s}, t)$ to denote a random field (i.e., spatiotemporal stochastic process) over $\mathcal{D}$; for purely spatial processes we drop the time index t. Gaussian stochastic processes are fully determined by the mean and the covariance kernel.

2.1 *The De Wijs Process*

The De Wijs process is defined in $d = 2$ by means of the following SPDE:

$$(-\triangle)^{1/2} X(\mathbf{s}) = \epsilon(\mathbf{s}) , \tag{1}$$

where $(-\triangle)^{1/2}$ is the half-Laplacian and $\epsilon(\mathbf{s})$ a Gaussian white noise random field [2, 9]. The De Wijs process is non-stationary and admits a generalized covariance which represents Green's function of the *Laplace equation*, namely, $\triangle C(\mathbf{r}) = -\delta(\mathbf{r})$, where $\mathbf{r}$ is the spatial lag. Hence, $C(\mathbf{r})$ can also be viewed as the solution of the 2D Poisson equation. The De Wijs process is characterized by the logarithmic variogram function $c_0 \ln(\|\mathbf{r}\|/r_s)$.

2.2 The Whittle-Matérn Process

The Whittle-Matérn SPDE is a fractional stochastic partial differential equation defined as

$$(\kappa^2 - \Delta)^{\frac{\nu}{2}} X(\mathbf{s}) = \epsilon(\mathbf{s}), \tag{2}$$

where $X(\mathbf{s})$ is the Whittle-Matérn random field, ν is the smoothness parameter, Δ is the Laplace operator, $\kappa = \sqrt{2\nu}/\xi$ is a ν-weighted inverse correlation scale, and $\epsilon(\mathbf{s})$ is a Gaussian white noise random field such that $\sigma(\mathbf{s}) \stackrel{d}{=} \mathcal{N}(0, \sigma_\epsilon^2)$, where $\sigma_\epsilon^2 = \sigma^2 \Gamma(\nu)(2\pi)^{-d/2}\kappa^{2\nu}$, σ^2 being the variance of the stationary Whittle-Matérn process. The respective Whittle-Matérn covariance kernel is given by the function

$$C(\mathbf{s} - \mathbf{s}') = \frac{2^{1-\nu}}{\Gamma(\nu)} \left(\kappa\sqrt{2\nu}\,\|\mathbf{s} - \mathbf{s}'\|\right)^{\nu} K_\nu(\kappa\sqrt{2\nu}\,\|\mathbf{s} - \mathbf{s}'\|), \tag{3}$$

where $\Gamma(\cdot)$ is the Gamma function and $K_\nu(\cdot)$ is the modified Bessel function of the second kind of order ν. Furthermore, the *spectral density* of the kernel is given by

$$S(\mathbf{k}) = \sigma^2 \frac{2^d \pi^{d/2} \Gamma(\nu + d/2)\, \xi^d}{\Gamma(\nu)\, \left(1 + \xi^2 \|\mathbf{k}\|^2\right)^{(\nu + d/2)}}. \tag{4}$$

The above equations are valid for any $d \in \mathbb{N}$, where $\mathbb{N}$ is the set of natural numbers.

2.3 Boltzmann-Gibbs Processes with Laplace-Polynomial Kernels

We will use the term "Laplace-polynomial" for SPDEs of the form

$$\Pi\left[(-\Delta)^{1/2}\right] X(\mathbf{s}) = \epsilon(\mathbf{s}),$$

where $\Pi(z) = \sum_{p=0}^{P} c_p\, z^p$ denotes the characteristic polynomial and $\epsilon(\mathbf{s}) \stackrel{d}{=} \mathcal{N}(0, \sigma_\epsilon^2)$. Note that the coefficients c_p of the characteristic polynomial must satisfy certain stability conditions to ensure stationarity of the process $X(\mathbf{s})$.

The De Wijs process is the simplest example in this class. The term "Boltzmann-Gibbs" is used because such processes admit joint probability density functions, f, which can be expressed in the Boltzmann-Gibbs formalism $f = \exp(-H)/Z$, where H is a quadratic "energy" functional of $X(\mathbf{s})$ and Z is a normalizing constant [1, 9, 10].

While the De Wijs process is nonstationary, higher-degree polynomials $\Pi(\cdot)$ can generate stationary covariance kernels. Herein we focus on the linear SPDE that generates *Spartan spatial random fields* [8]; in spaces with dimension $d \leq 3$ it is given by the following equation [9]:

$$\frac{1}{\sqrt{\eta_0 \xi^d}} \left[1 - \left(\sqrt{2+\eta_1}\right) \xi \, (-\Delta)^{1/2} + \xi^2 \Delta \right] X(\mathbf{s}) = \epsilon(\mathbf{s}), \tag{5}$$

where $\eta_0 > 0$ is a scale factor, $\eta_1 > -2$ is the rigidity coefficient, and $\xi > 0$ is the characteristic length.

Using (i) the fact that Π is a linear superposition of integer powers of the half-Laplacian $(-\Delta)^{1/2}$ and (ii) that for any weakly differentiable function Φ, $(-\Delta)_{\mathbf{s}}^{1/2} \Phi(\mathbf{s}-\mathbf{s}') = -(-\Delta)_{\mathbf{s}'}^{1/2} \Phi(\mathbf{s}-\mathbf{s}')$, the Spartan covariance kernel (assuming an infinite domain) respects the following partial differential equation (PDE):

$$\mathcal{L}\, C(\mathbf{s}-\mathbf{s}') = \delta(\mathbf{s}-\mathbf{s}'), \tag{6}$$

where $\mathcal{L}$ is the *self-adjoint SSRF precision* operator defined by

$$\mathcal{L} = \frac{1}{\eta_0 \xi^d} \left[1 - \eta_1 \xi^2 \Delta_{\mathbf{s}} + \xi^4 \Delta_{\mathbf{s}}^2 \right]. \tag{7}$$

Note that if the characteristic polynomial Π is of degree-p, then the precision operator $\mathcal{L}$ involves a polynomial of degree p with respect to the Laplacian.

According to (6), the covariance kernel is *Green's function* of the homogeneous PDE $\mathcal{L}\,\Upsilon(\mathbf{s}) = 0$ on an infinite domain. The correspondence between covariance functions resulting from SPDEs and Green functions of associated PDEs is well known in the physics literature [4]; however, it had not been explored until [8] for the generation of spatial covariance kernels.

Based on (6) and (7), it can be shown by transforming $\mathcal{L}$, $C(\cdot)$, $\delta(\cdot)$ in the Fourier space that the spectral density of SSRFs is given by [8, 9]

$$S(\mathbf{k}) = \frac{\eta_0 \xi^d}{1 + \eta_1 \, (\|\mathbf{k}\|\xi)^2 + (\|\mathbf{k}\|\xi)^4}. \tag{8}$$

Is there a physical motivation for the particular form of the SPDE (5)? It turns out that in $d = 1$, the covariance equation (7) describes a *stochastic oscillator* driven by white noise [9, 11]. This is a paradigmatic physical model, also known as Brownian oscillator [21]. Hence, in $d = 2, 3$ the respective covariance kernels can be viewed as extensions of the one-dimensional harmonic oscillator.

The functional form of the SSRF covariance kernel depends on $d \in \{1, 2, 3\}$ and the rigidity coefficient η_1: there exist three different regimes for $|\eta_1| < 2$, for $\eta_1 = 2$, and for $\eta_1 > 2$. The respective SSRF kernel expressions, given below, comprise three different regimes depending on the value of η_1 [9, 12].

In one dimension a slightly different parametric form is used (to comply with standard harmonic oscillator notation): τ represents the time lag, ω_d the oscillation frequency, and τ_c the relaxation time.

$$C(\tau)=\frac{\sigma^2\, e^{-\frac{|\tau|}{2\tau_c}}}{2\omega_0^2\,\tau_c}\left(\cos\omega_d\tau+\frac{\sin\omega_d|\tau|}{2\omega_d\tau_c}\right),\ \omega_0\tau_c>1/2,\qquad \text{Underdamping} \tag{9a}$$

$$C(\tau)=\frac{\sigma^2\, e^{-\frac{|\tau|}{2\tau_c}}}{2\omega_0^2\,\tau_c}\left(1+\frac{|\tau|}{2\tau_c}\right),\ \omega_0\tau_c=1/2,\qquad \text{Critical damping} \tag{9b}$$

$$C(\tau)=\frac{\sigma^2}{4\omega_0^2\,\tau_c}\left(\frac{e^{-|\tau|/\tau_s}}{|\omega_d|\tau_f}-\frac{e^{-|\tau|/\tau_f}}{|\omega_d|\tau_s}\right),\ \omega_0\tau_c<1/2\,.\qquad \text{Overdamping} \tag{9c}$$

In $d=2$, using the normalized lag $h=\|\mathbf{r}\|/\xi$, the SSRF kernel is expressed as follows:

$$C(h)=\frac{\eta_0\,\mathrm{Im}\,[K_0(h\,z_+)]}{\pi\sqrt{4-\eta_1{}^2}},\quad |\eta_1|<2\,, \tag{10a}$$

$$C(h)=\left(\frac{\eta_0\,h}{4\pi}\right)\,K_{-1}(h),\quad \eta_1=2\,, \tag{10b}$$

$$C(h)=\frac{\eta_0\,[K_0(h\,z_+)-K_0(h\,z_-)]}{2\pi\sqrt{\eta_1{}^2-4}},\quad \eta_1>2\,, \tag{10c}$$

where Im denotes the imaginary part, $z_\pm=\sqrt{-t_\pm^*}$, and $t_\pm^*=\left(-\eta_1\pm\sqrt{\eta_1{}^2-4}\right)/2$, while $K_\nu(z)$ is the modified Bessel function of the second kind and order ν.

Finally, in $d=3$, the SSRF kernel is dominated by exponential dependence, and it is expressed as follows in the three η_1 regimes:

$$C(h)=\eta_0\,\frac{e^{-h\beta_2}}{2\,\pi\,\Delta}\left[\frac{\sin\,(h\beta_1)}{h}\right],\quad |\eta_1|<2\,, \tag{11a}$$

$$C(h)=\frac{\eta_0}{8\,\pi}\,e^{-h},\quad \eta_1=2\,, \tag{11b}$$

$$C(h)=\frac{1}{4\,\pi\,\Delta}\left[\frac{e^{-h\omega_1}-e^{-h\omega_2}}{h\,(\omega_2-\omega_1)}\right],\quad \eta_1>2\,, \tag{11c}$$

where $\beta_{1,2}=\left(\frac{|2\mp\eta_1|}{4}\right)^{1/2}$, $\omega_{1,2}=\left(\frac{|\eta_1\mp\Delta|}{2}\right)^{1/2}$, and $\Delta=|\eta_1{}^2-4|^{\frac{1}{2}}$.

Spartan random fields were the first Boltzmann-Gibbs model derived from a second-degree Laplace-polynomial SPDE. Other models are also possible using the same approach with Laplace polynomials of different degrees, as described in [1, 9, 10, 22].

2.4 Comparing Matérn and Spartan Random Fields

Matérn and Spartan covariance kernels are both generated by SPDEs. However, there are significant differences between the two families.

1. The Matérn kernels have a functional form in real space which is independent of d, while their spectral density depends on d. In SSRFs, the spectral density is independent of d in contrast with the real-space kernels.
2. Isotropic Matérn and SSRF models involve three parameters: the variance and characteristic length scale are shared by both; Matérn models also involve the smoothness parameter ν, while SSRFs include the rigidity η_1.
3. In Matérn models, ν controls the differentiability of the underlying field (higher ν values imply smoother spatial variations). In SSRFs, η_1 controls the kernel shape: for $|\eta_1| < 2$, the respective kernels exhibit a negative hole effect (this is more pronounced for lower d). On the other hand, Matérn kernels decay monotonically towards zero.
4. The smoothness of Matérn kernels is determined by ν. SSRFs are once differentiable in $d = 1$ and continuous but non-differentiable (in the mean-square sense) for $d = 2, 3$ [12]. To obtain kernels that admit higher-order mean-square derivatives, higher-degree polynomials of the Laplacian should be used in the precision operator $\mathcal{L}$ in Eq. (7).
5. SSRFs can be expressed in terms of Boltzmann-Gibbs probability density functions with sparse interactions (i.e., short-range precision operator) in the respective energy functional. This property offers tangible computational advantages, as it eliminates the need to invert covariance matrices [9, 10].

3 Space-Time Covariance Kernels from SPDEs?

As discussed above, for purely spatial problems, stationary kernels (at least for Euclidean distance measures) can be derived directly from SPDEs. However, the problem is considerably more difficult for space-time domains. Non-separable space-time covariance kernels, including the Gneiting class [6, 14, 17], exist. These models are often based on mathematical principles that ensure positive-definiteness, but they do not correspond to solutions of physically inspired PDEs.

To our knowledge, the only available non-separable space-time kernel that is derived by solving a PDE is the space-time covariance function developed by Heine [7]. The Heine kernel represents the solution of a parabolic PDE with one temporal and one spatial dimension.

A different approach based on the linear response theory of statistical mechanics is used in [13]. The linear response of a space-time field $X(\mathrm{s}, t)$ around an equilibrium state, $X(\mathbf{s})$, is given by

$$\frac{\partial X(\mathbf{s},t)}{\partial t} = -\Gamma \left. \frac{\delta H[X(\mathbf{s});\boldsymbol{\theta}]}{\delta X(\mathbf{s})} \right|_{X(\mathbf{s})=X(\mathbf{s},t)} + \epsilon(\mathbf{s},t), \tag{12}$$

where $\Gamma > 0$ is a diffusion coefficient that determines the relaxation towards the equilibrium state, $H[x(\mathbf{s});\boldsymbol{\theta}]$ is the energy functional at equilibrium, $\delta(\cdot)/\delta X(\mathbf{s})$ is the functional derivative with respect to the field $X(\mathbf{s})$, and $\epsilon(\mathbf{s},t)$ is the Gaussian white noise field. By inserting the SSRF energy functional H in Eq. (12), a space-time PDE (with fourth-order spatial derivatives) is obtained for the covariance kernel. This PDE can be solved in $d = 1$, recovering the Heine solution

$$C_1(h,u;\boldsymbol{\theta}) = \frac{\eta_0 \lambda}{4} \left[e^{-\lambda h} \operatorname{erfc}\left(\sqrt{u} - \frac{\lambda h}{2\sqrt{u}}\right) + e^{\lambda h} \operatorname{erfc}\left(\sqrt{u} + \frac{\lambda h}{2\sqrt{u}}\right)\right], \tag{13}$$

where $h = \|r\|/\xi$ and $u = |\tau|/\tau_c$ are the normalized space and time lags, and $\boldsymbol{\theta} = (\eta_0, \lambda, \xi, \tau_c)^\top$ is the space-time parameter vector. The *flexibility (inverse rigidity)* constant is defined as $\lambda = 1/\sqrt{\eta_1}$, and $\operatorname{erfc}(\cdot)$ is the complementary error function defined by the integral

$$\operatorname{erfc}(x) = \frac{2}{\sqrt{\pi}} \int_x^\infty dt\, e^{-t^2}.$$

The linear response model has not yielded explicit solutions in spatial dimensions higher than $d = 1$. Application of the turning bands transform to the $1+1$ space-time covariance (13) leads to a non-separable space-time covariance kernel which is permissible for $d = 3$ [13]. However, this kernel does not represent a PDE solution in $d + 1$ dimensions.

4 On the Hybrid Spectral Approach

A hybrid spectral approach (HSA) for covariance kernel generation has been developed, leveraging physical arguments [11]. The main ideas underlying HSA are summarized below:

1. The first step of HSA is the development of a temporal correlation kernel by explicitly solving Green's functions that are associated with stochastic ordinary differential equations (SODEs). Typical examples involve the first-order *Ornstein-Uhlenbeck* equation which represents noisy relaxation to a steady state and the second-order Brownian oscillator equation which can capture noisy oscillatory patterns. In the latter case, the covariance kernel is given by Eq. (9).
2. The second step of HSA injects spectral dependence into the temporal kernel via the wavevector $\mathbf{k}$. This is accomplished by replacing constant SODE coefficients with frequency-dependent functions. The $\mathbf{k}$-dependence is selected so that

the coefficients respect physical constraints (e.g., nonnegativity or meaningful dispersion relations). In addition, it must be established that the resulting **k**-modulated function is a valid spectral density (i.e., it satisfies Bochner's theorem) for all time lags $\tau \in \mathbb{R}$.

3. The last step requires explicit integration of the **k** dependence (via the inverse Fourier transform) which, in certain cases, leads to closed-form covariance kernel expressions in $d + 1$ dimensions with inherent space-time interactions.

A detailed presentation of the above mathematical steps is given in [11], where covariance kernel expressions are derived for the stochastic oscillator and the stochastic relaxation (Ornstein-Uhlenbeck) models. Below we focus on the simpler Ornstein-Uhlenbeck model, aiming to clarify the steps of the HSA procedure.

4.1 *Ornstein-Uhlenbeck Process*

The Ornstein-Uhlenbeck SODE is given by

$$\frac{dz(t;\zeta)}{dt} + \frac{1}{\tau_c} z(t;\zeta) = \sigma_\eta \eta(t;\zeta), \tag{14}$$

where $\eta(t;\zeta) \sim \mathcal{N}(0,1)$ and $\sigma_\eta > 0$ is the white noise standard deviation. The covariance of the O-U process is given by the exponential function [16, p. 448]

$$C(\tau) = \sigma^2 \exp(-|\tau|/\tau_c),$$

where $\sigma^2 = \sigma_\eta^2 \tau_c/2$ and τ_c is the relaxation time. In HSA, the model parameters are "dressed" with wavevector dependence which ultimately leads to spatial correlations. The **k** dependence is introduced by means of dispersion functions.

The *dispersion relations* that insert **k** dependence in the parameters of the O-U covariance kernel are given by the following equations:

$$\tau_c \to \tau_c(\mathbf{k}) = \tilde{\tau}_c/B(\mathbf{k}), \quad \sigma^2 \to \sigma^2(\mathbf{k}) = \sigma_0^2 A(\mathbf{k}), \tag{15}$$

where $\tilde{\tau}_c$ and σ_0^2 are positive constants. While the dispersion functions in general depend on **k**, and thus they could incorporate anisotropy; in the following we focus on radially symmetric dispersion functions $A(k)$ and $B(k)$, where $k = ||\mathbf{k}||$ is the norm of the wavevector **k**, which is also known as the spatial frequency.

Physically, the above equations imply that different modes, corresponding to different **k**, have their own variance and relaxation time as determined by the dispersion functions $A(k)$ and $B(k)$. We can reasonably assume that (i) the mode variance should decrease with increasing k—since opposite behavior can lead to explosive growth of the variance, a phenomenon known in physics as ultraviolet divergence—and (ii) the mode relaxation time should also decrease with increasing k.

In general, relaxation times in physical phenomena decrease with increasing mode frequency. For example, higher-frequency modes may correspond to more energetic and rapid oscillations, which tend to dissipate energy faster due to interactions with surrounding particles or via damping mechanisms. In microscopic phenomena—such as processes that involve phonons in solids or molecular vibrations—higher-frequency modes often experience shorter lifetimes due to decay processes, such as anharmonic interactions, or coupling to other excitations. This means that the dispersion function $B(k)$ should be an increasing function of k in order to reduce the relaxation time for higher frequencies according to Eq. (15), while $A(k)$ should be a decreasing function of k in order to reduce the variance of high-frequency modes.

The above constraints allow considerable flexibility regarding the behavior of the dispersion functions. In addition, we have postulated monotonic dependence of the functions $A(k)$ and $B(k)$, but this could be unnecessarily restrictive. The main requirement is that the modal relaxation time and variance decrease asymptotically. However, there could be an intermediate frequency range where these dispersion functions behave non-monotonically. Hence, more than one spatiotemporal covariance kernels can be derived from the temporal Ornstein-Uhlenbeck covariance kernel. In [11] two families of space-time covariance kernels were obtained.

(1) Quadratic dependence of dispersion functions: By setting $A(k) = e^{-\beta k^2}$, $B(k) = a + bk^2$, where the constants $a, b, \beta > 0$ are parameters with units $[b] = [\beta] = [L]^2$, $[a] = [L]^0$ ($[L]$ representing units of length), the following spatiotemporal kernel is obtained by spectral integration in d dimensions:

$$C(r,\tau) = \frac{\sigma_0^2\, e^{-a\,|\tau|/\tilde{\tau}_c}}{(2\pi)^{d/2}} \; \frac{e^{-r^2/4(\beta + b\,|\tau|/\tilde{\tau}_c)}}{\left(\frac{2b|\tau|}{\tilde{\tau}_c} + 2\beta\right)^{d/2}}. \tag{16}$$

The kernel (16) involves four free parameters: the standard deviation, σ_0, the time constant, $\tilde{\tau}_c/a$, the spatial scale, $2\sqrt{\beta}$, and the space-time interaction parameter b.

(2) Linear dependence of dispersion functions: If we set $A(k) = e^{-\beta k}$, $B(k) = a + \xi\, k$, where $a, \xi, \beta > 0$ are parameters with units $[\xi] = [\beta] = [L]$, $[a] = [L]^0$, the following space-time kernel function is obtained:

$$C(r,\tau) = \frac{\sigma_0^2\, \Gamma(\frac{d+1}{2})}{\pi^{(d+1)/2}} \; \frac{(\beta\tilde{\tau}_c + \xi\,|\tau|)\; e^{-a\,|\tau|/\tilde{\tau}_c}}{\tilde{\tau}_c \left[r^2 + \left(\beta + \frac{\xi|\tau|}{\tilde{\tau}_c}\right)^2\right]^{(d+1)/2}}, \tag{17}$$

where $\Gamma(\cdot)$ is the Gamma function. The kernel (17) also involves four parameters, namely the standard deviation, σ_0, the time constant, $\tilde{\tau}_c/a$, the spatial scale, β, and the space-time interaction parameter, ξ (or $\xi/\tilde{\tau}_c$).

For both O-U kernels, the role of the parameters is clarified if one considers the marginal kernels $C(r, \tau = 0)$ and $C(r = 0, \tau)$, the explicit expressions for which are given in [11].

5 Conclusions

The formulation of physically inspired covariance kernels has always attracted attention in the communities of geostatistics, space-time statistics, and machine learning. In particular, it is desired to have space-time kernels that can be associated with differential equations that represent physical processes. As discussed herein, this goal has been achieved to some extent in the purely spatial case. Characteristic examples involve the Matérn family and the cases of Boltzmann-Gibbs covariance kernels with precision operators that represent polynomials of the Laplace operator. However, such constructions cannot be easily extended to the space-time domain. To our knowledge, the only space-time kernel that is associated with a stochastic differential equation and admits explicit expression is the Heine model, which is valid in an idealized space-time domain with one spatial dimension.

Motivated by the lack of closed-form, space-time covariance kernels derivable from partial differential equations, we recently proposed the hybrid spectral approach [11]. As explained herein, the main idea behind HSA is that the temporal behavior of the kernel can be derived by analytical solution of some ODE. Then, we can insert spatial correlations in the kernel by introducing frequency (wavevector) dependence of the temporal kernel's parameters. Furthermore, we argued that the wavevector dependence can be introduced by meaningful dispersion functions, which respect constraints observed in physical systems. If the dispersion relations allow for explicit integration of the spectral dependence, it is possible to obtain closed-form space-time kernel expressions as described in [11]. Hence, the HSA provides an indirect approach of constructing non-separable space-time kernels that respect physical equations of motion in the time domain (e.g., noisy relaxation and stochastic oscillations) and physically motivated constraints for the dispersion of the temporal coefficients.

Bibliography

1. Allard, D., Hristopulos, D.T., Opitz, T.: Linking physics and spatial statistics: a new family of Boltzmann-Gibbs random fields. Electron. J. Stat. **15**(2), 4085–4116 (2021)
2. Besag, J., Mondal, D.: First-order intrinsic autoregressions and the de Wijs process. Biometrika **92**(4), 909–920 (2005)
3. Christakos, G., Hristopulos, D.T.: Spatiotemporal Environmental Health Modelling: A Tractatus Stochasticus. Kluwer Academic, Dordrecht (1997)
4. Dolph, C.L., Woodbury, M.A.: On the relation between Green's functions and covariances of certain stochastic processes and its application to unbiased linear prediction. Trans. Am. Math. Soc. **72**(3), 519–550 (1952)
5. Genton, M.G.: Separable approximations of space-time covariance matrices. Environmetrics **18**(7), 681–695 (2007)
6. Gneiting, T.: Nonseparable, stationary covariance functions for space-time data. J. Am. Stat. Assoc. **97**(458), 590–600 (2002)
7. Heine, V.: Models for two-dimensional stationary stochastic processes. Biometrika **42**(1–2), 170–178 (1955)

8. Hristopulos, D.T.: Spartan Gibbs random field models for geostatistical applications. SIAM J. Scient. Comput. **24**(6), 2125–2162 (2003)
9. Hristopulos, D.T.: Random Fields for Spatial Data Modeling: A Primer for Scientists and Engineers. Springer, Dordrecht (2020)
10. Hristopulos, D.T.: Boltzmann–Gibbs random fields with mesh-free precision operators based on smoothed particle hydrodynamics. Theory Probab. Math. Statist. **107**, 37–60 (2022)
11. Hristopulos, D.T.: Non-separable covariance kernels for spatiotemporal Gaussian processes based on a hybrid spectral method and the harmonic oscillator. IEEE Trans. Inf. Theory **70**(2), 1268–1283 (2024)
12. Hristopulos, D.T., Elogne, S.N.: Analytic properties and covariance functions for a new class of generalized Gibbs random fields. IEEE Trans. Inf. Theory **53**(12), 4667–4679 (2007)
13. Hristopulos, D.T., Tsantili, I.: Space-time covariance functions based on linear response theory and the turning bands method. Spatial Stat. **22**(Part 2), 321–337 (2017)
14. Kolovos, A., Christakos, G., Hristopulos, D.T., Serre, M.L.: Methods for generating non-separable spatiotemporal covariance models with potential environmental applications. Adv. Water Resources **27**(8), 815–830 (2004)
15. Lindgren, F., Rue, H., Lindström, J.: An explicit link between Gaussian fields and Gaussian Markov random fields: the stochastic partial differential equation approach. J. R. Stat. Soc. Ser. B **73**(4), 423–498 (2011)
16. Papoulis, A., Pillai, S.U.: Probability, random variables and stochastic processes. Series in Electrical and Computer Engineering, 4 edn. McGraw-Hill, New York (2002)
17. Porcu, E., Furrer, R., Nychka, D.: 30 years of space-time covariance functions. Wiley Interdisciplinary Reviews. Computat. Stat. **13**(2), e1512 (2021)
18. Rougier, J.: A representation theorem for stochastic processes with separable covariance functions, and its implications for emulation. arXiv preprint. arXiv:1702.05599 (2017)
19. Taira, K., et al.: Modal analysis of fluid flows: an overview. Am. Instit. Aeronaut. Astronaut. J. **55**, 4013–4041 (2017)
20. Tu, J.H., Rowley, C.W., Luchtenburg, D.M., Brunton, S.L., Kutz, J.N.: On dynamic mode decomposition: theory and applications. J. Comput. Dyn. **1**(2), 391–421 (2014)
21. Uhlenbeck, G.E., Ornstein, L.S.: On the theory of the Brownian motion. Phys. Rev. **36**, 823–841 (1930)
22. Yaremchuk, M., Smith, S.: On the correlation functions associated with polynomials of the diffusion operator. Q. J. R. Meteorol. Soc. **137**(660), 1927–1932 (2011)

Stationary Subspace Analysis for Spatiotemporal Data

Jaakko Pere, Sandra De Iaco, and Klaus Nordhausen

Abstract Multivariate spatiotemporal data, where multiple variables are measured over time and across space, present modeling challenges due to complex dependencies. Traditional approaches often rely on assumptions like stationarity and separability, which may be too restrictive. This work introduces a flexible alternative by decomposing the random field into two independent subspaces: one stationary and one nonstationary. The proposed method, spatiotemporal stationary subspace analysis (stSSA), identifies the subspaces by segmenting the space-time domain and comparing local and global means. This decomposition simplifies modeling and enhances interpretability. A simulation study assesses the method's sensitivity to segment construction, number of latent components and the choice of the covariance function.

Keywords Blind source separation · Dimension reduction · Non-stationarity

1 Introduction

Increasingly, many variables are observed at different locations and regularly tracked over time, yielding so-called multivariate spatiotemporal data. Such data are challenging to model, as dependencies between the variables in space and time must be considered. The classical approach is to model such data using a p-variate random field $\mathbf{x} = \mathbf{x}(\mathbf{u}, t)$, where $(\mathbf{u}, t) \in \mathcal{U} \times \mathcal{T} \subset \mathbb{R}^{d+1}$, with $\mathbf{u} \in \mathcal{U} \subset \mathbb{R}^d$

J. Pere · K. Nordhausen (✉)
University of Helsinki, Helsinki, Finland
e-mail: klaus.nordhausen@helsinki.fi

S. De Iaco
Department of Economic Sciences, University of Salento, Lecce, Italy

Big Data and Quantum Computing, National Centre for HPC, Bologna, Italy

National Biodiversity Future Center, Palermo, Italy

S. De Iaco et al. (eds.), *Exploration of Spatio-Temporal Environmental Conditions: Harmonized Databases and Analytical Techniques*, Springer Proceedings in Mathematics & Statistics 531, https://doi.org/10.1007/978-3-032-17526-7_4

denoting the spatial domain (typically $d \in \{1, 2, 3\}$) and $\mathcal{T} \subset \mathbb{R}$ denoting the temporal domain.

Given the complex dependency structures in such data, it is common to make simplifying assumptions about the first two moments. A standard assumption is weak stationarity, which implies that $\mathbb{E}(\mathbf{x}(\mathbf{u}, t)) = \boldsymbol{\mu} \in \mathbb{R}^p$ for all $(\mathbf{u}, t) \in \mathcal{U} \times \mathcal{T}$, meaning that the expectation is constant across the entire domain and that the matrix-variate covariance function depends only on the spatial and temporal distances, i.e., the covariance can be expressed as $\operatorname{Cov}\left(\mathbf{x}(\mathbf{u}, t), \mathbf{x}(\mathbf{u}', t')\right) = C\left(\mathbf{u} - \mathbf{u}', t - t'\right)$, for some suitable function $C : \mathbb{R}^d \times \mathbb{R} \to \mathbb{R}^{p \times p}$. Stronger assumptions are also possible, such as isotropy (in space) or separability of spatial and temporal dependencies [2, 9].

However, such simplifying assumptions often come at a cost, like inferior prediction performance. Hence, there has been growing interest in nonstationary approaches [1, 10, 12]. These nonstationary approaches, though flexible, are often complex and tend to scale poorly with increasing data dimension p. Additionally, they usually assume non-stationarity across all variables. Similarly, dimension reduction methods used as a preprocessing step often require either all components to be stationary [3, 8] or all to be nonstationary [11].

We are interested in a different form of dimension reduction for the field $\mathbf{x}$, namely, in one that decomposes it into stationary and nonstationary parts, which are independent of each other, after which the two independent random fields can be modeled separately. This dimension reduction approach is known in a time series context as stationary subspace analysis (SSA) [4], and we will refer to it here as spatiotemporal SSA (stSSA). Our focus will be on data where non-stationarity is present in the mean, while other forms of non-stationarity will be discussed in future work. Note that the approach considered here differs from the one where a random field is decomposed into a sum or product of stationary and nonstationary parts as discussed in [5].

The remainder of this chapter is organized as follows. Section 2 presents the assumptions of the stSSA model, and Sect. 3 describes the developed method for the estimation of the stationary and nonstationary subspaces. Section 4 gives a simulation study that demonstrates the method. Lastly, Sect. 5 concludes the paper by summarizing the contributions and offering future research directions.

2 Model Formulation

The idea of the stSSA model is that the underlying latent fields can be divided into stationary and nonstationary ones. Below, we present the stSSA model, which is an extension of the SSA model [4] to a spatiotemporal setting.

Definition 1 A p-variate random field $\mathbf{x} = \mathbf{x}(\mathbf{u}, t) = \left(x_1(\mathbf{u}, t), \ldots, x_p(\mathbf{u}, t)\right)^{\top}$ on the domain $\mathcal{U} \times \mathcal{T} \subset \mathbb{R}^{d+1}$ follows the stSSA model if

$$\mathbf{x}(\mathbf{u},t) = \mathbf{A}\mathbf{z}(\mathbf{u},t) = \left[\mathbf{A_s}\ \mathbf{A_n}\right]\begin{pmatrix}\mathbf{s}(\mathbf{u},t)\\ \mathbf{n}(\mathbf{u},t)\end{pmatrix},$$

where $\mathbf{A} = \left[\mathbf{A_s}\ \mathbf{A_n}\right] \in \mathbb{R}^{p\times p}$ is full-rank not depending on location $\mathbf{u}$ or time t, $\mathbf{A_s} \in \mathbb{R}^{p\times(p-q)}$, $\mathbf{A_n} \in \mathbb{R}^{p\times q}$. Let $\mathbf{I}_r$ denote the $r \times r$ identity matrix, and let $\mathbf{D}_r^{(\mathbf{u},t)}$ be an $r \times r$ diagonal matrix with positive diagonal elements, possibly depending on the location $\mathbf{u}$ and time t. The p-variate random field $\mathbf{z}$, consisting of $p - q$-variate and q-variate random fields $\mathbf{s}$ and $\mathbf{n}$, satisfies the following assumptions:

(A1) $\mathbb{E}\left(\mathbf{s}(\mathbf{u},t)\right) = \mathbf{0}$ and $\mathrm{Cov}(\mathbf{s}(\mathbf{u},t)) = \mathbf{I}_{p-q}$ for all $(\mathbf{u},t) \in \mathcal{U} \times \mathcal{T}$, and $\mathrm{Cov}\left(\mathbf{s}(\mathbf{u},t), \mathbf{s}(\mathbf{u}',t')\right)$ is finite and only a function of $\|\mathbf{u}-\mathbf{u}'\|$ and $\|t-t'\|$.

(A2) $\mathbb{E}(\mathbf{n}(\mathbf{u},t)) < \infty$ and $\mathrm{Cov}(\mathbf{n}(\mathbf{u},t)) = \mathbf{D}_q^{(\mathbf{u},t)}$ for all $(\mathbf{u},t) \in \mathcal{U} \times \mathcal{T}$, and $\mathrm{Cov}\left(\mathbf{n}(\mathbf{u},t), \mathbf{n}(\mathbf{u}',t')\right)$ is finite.

(A3) The random fields $\mathbf{s}$ and $\mathbf{n}$ are independent, from which it follows that $\mathrm{Cov}(\mathbf{s}(\mathbf{u},t), \mathbf{n}(\mathbf{u}',t')) = \mathbf{0}$ for all $(\mathbf{u},t), (\mathbf{u}',t') \in \mathcal{U} \times \mathcal{T}$.

(A4) The dimension q is the smallest value that allows division of $\mathbf{x}$ such that Assumptions (A1)–(A3) hold.

(A5) There exist $U_0 = \{\mathbf{u}_1, \ldots, \mathbf{u}_{n_0}\} \subset \mathcal{U}$ and $[T] = \{1, \ldots, T\} \subset \mathcal{T}$ such that $\sum_{(\mathbf{u},t)\in U_0\times[T]} \mathbb{E}(\mathbf{n}(\mathbf{u},t)) = \mathbf{0}$ and $\sum_{(\mathbf{u},t)\in U_0\times[T]} \mathrm{Cov}(\mathbf{n}(\mathbf{u},t)) = \mathbf{I}_q$.

By Assumption (A1), the $(p-q)$ random fields in $\mathbf{s}$ are second-order stationary with finite second-order spatiotemporal dependence, and additionally, the location and covariance matrix of the stationary part are fixed for convenience. On the other hand, Assumption (A2) states that the q nonstationary components in $\mathbf{n}$ have finite first and second moments and are uncorrelated at $(\mathbf{u},t)$ but can have arbitrary spatiotemporal second-order dependence for $(\mathbf{u},t) \neq (\mathbf{u}',t')$. Assumption (A3) states that the nonstationary and stationary components are independent, and Assumption (A4) ensures that all the components of $\mathbf{n}$ are nonstationary.

Despite Assumptions (A1)–(A4) the model is not well defined. The stationary components are only specified up to a rotation by an orthogonal matrix. Furthermore, the nonstationary components can be marginally rescaled, shifted, and permuted. Thus, we have Assumption (A5) fixing the location and the scale of $\mathbf{n}$ based on a set of reference points $U_0 \times [T]$. Additionally, we will assume that the observed sample $\mathbf{x}\left(\mathbf{u}_i, t_j\right)$, $i \in \{1, \ldots, n_s\}$, $j \in \{1, \ldots, n_t\}$, can always be chosen to be the set of reference points such that Assumption (A5) is satisfied.

Throughout, we assume that the field $\mathbf{x}$ follows the stSSA model in Definition 1. The goal is to estimate the unmixing matrix $\mathbf{W}$ so that $\mathbf{W}\mathbf{x}(\mathbf{u},t)$ can be partitioned into stationary and nonstationary fields. However, notice that the unmixing matrix is not uniquely defined since, for example, columns of $\mathbf{A}_s$ and $\mathbf{A}_n$ can be permutated. This has to be taken into account while measuring the performance of the estimation. Both the assumed model for $\mathbf{x}$ and the method for estimating $\mathbf{W}$ are called stSSA. It should be clear from the context if we refer to the model or the method of estimating $\mathbf{W}$.

3 The stSSA Method for the Mean

In the present chapter, we restrict to detecting non-stationarity in the mean. For example, the nonstationary components $n_i = n_i(\mathbf{u}, t)$ of $\mathbf{n} = (n_1, \ldots, n_q)$ can exhibit trend, seasonality, or cyclicity.

Suppose that the field $\mathbf{x}$ is observed at the locations $\{\mathbf{u}_i\}_{i=1}^{n_s}$ and time points $\{t_j\}_{j=1}^{n_t}$. The spatiotemporal locations $(\mathbf{u}_i, t_j)$ are divided into non-overlapping segments [11]. That is, firstly, the space-time domain $\mathcal{U} \times \mathcal{T}$ is divided into K parts by a partition $\mathcal{S}_1 = \mathcal{U}_1 \times \mathcal{T}_1, \ldots, \mathcal{S}_K = \mathcal{U}_K \times \mathcal{T}_K$ such that $\mathcal{S}_i \cap \mathcal{S}_j = \varnothing$ for $1 \leq i \neq j \leq K$ and $\bigcup_{k \in K} \mathcal{S}_k = \mathcal{U} \times \mathcal{T}$. Secondly, an observation $\mathbf{x}(\mathbf{u}_i, t_j)$ is assigned to the segment $\mathbf{S}_k$ if $(\mathbf{u}_i, t_j) \in \mathcal{S}_k$, and let $|\mathbf{S}_k|$ denote the number of observations in $\mathbf{S}_k$. We denote $\mathbf{S} = \bigcup_{k \in K} \mathbf{S}_k$, and accordingly, $|\mathbf{S}|$ corresponds to the number of all realized space-time observations.

Based on the chosen space-time partition, the sample mean of each segment $\mathbf{S}_k$ is compared to the global sample mean. Denote the sample mean of a segment $\mathbf{S}_k$ by $\mathbf{m}_{\mathbf{S}_k}(\mathbf{x}) = |\mathbf{S}_k|^{-1} \sum_{(\mathbf{u}_i, t_j) \in \mathbf{S}_k} \mathbf{x}(\mathbf{u}_i, t_j)$. Hence, $\mathbf{m}_{\mathbf{S}}(\mathbf{x})$ denotes the global sample mean based on the whole sample of $\mathbf{x}$. Furthermore, we work with the whitened observations of the field $\mathbf{x}$, $\mathbf{x}^{st}(\mathbf{u}_i, t_j) = \mathbf{Q}(\mathbf{x})^{-1/2} \left(\mathbf{x}(\mathbf{u}_i, t_j) - \mathbf{m}_{\mathbf{S}}(\mathbf{x})\right)$, where $\mathbf{Q}_{\mathbf{S}}(\mathbf{x}) = |\mathbf{S}|^{-1} \sum_{(\mathbf{u}_i, t_j) \in \mathbf{S}} \left(\mathbf{x}(\mathbf{u}_i, t_j) - \mathbf{m}_{\mathbf{S}}(\mathbf{x})\right) \left(\mathbf{x}(\mathbf{u}_i, t_j) - \mathbf{m}_{\mathbf{S}}(\mathbf{x})\right)^\top$ is the sample covariance corresponding to the available sample of the field $\mathbf{x}$, and $\mathbf{A}^{-1/2}$ is the unique matrix corresponding to a symmetric positive definite matrix $\mathbf{A}$ such that $\mathbf{A}^{-1} = \mathbf{A}^{-1/2}\mathbf{A}^{-1/2}$. Now, for the whitened observations, segment sample means are compared to the global sample means by computing

$$\mathbf{M} = \sum_{k=1}^{K} \frac{|\mathbf{S}_k|}{|\mathbf{S}|} \mathbf{m}_{\mathbf{S}_k}(\mathbf{x}^{st}) \mathbf{m}_{\mathbf{S}_k}(\mathbf{x}^{st})^\top. \tag{1}$$

That is, $\mathbf{M} \in \mathbb{R}^{p \times p}$ is just the sample covariance matrix of sample means corresponding to different segments weighted by their number of observations.

The mean stationary components of the latent field $\mathbf{z}$ correspond to the eigenvalues of $\mathbf{M}$ that are close to zero. Likewise, the nonstationary components with respect to trend correspond to the nonzero eigenvalues of $\mathbf{M}$. Hence, the estimates for the stationary and nonstationary parts of the unmixing matrix

$$\mathbf{W} = \begin{pmatrix} \mathbf{W}_s \\ \mathbf{W}_n \end{pmatrix}, \quad \mathbf{W}_s \in \mathbb{R}^{(p-q) \times p} \quad \text{and} \quad \mathbf{W}_n \in \mathbb{R}^{q \times p} \tag{2}$$

are obtained through the eigenvalue decomposition of $\mathbf{M}$. Below, we give the definitions for the estimators of $\mathbf{W}_s$ and $\mathbf{W}_n$.

Definition 2 Let $\mathbf{x}$ be a random field following the stSSA model of Definition 1. Write the eigenvalue decomposition of the matrix $\mathbf{M}$ from Eq. (1) as $\mathbf{M} = \mathbf{V}\mathbf{D}\mathbf{V}^\top$, where $\mathbf{D}$ is a diagonal matrix with eigenvalues ordered by $\lambda_1 \geq \ldots \geq \lambda_p$, the ith column of $\mathbf{V} = \left[\mathbf{V}_n \ \mathbf{V}_s\right]$ is the eigenvector corresponding to the eigenvalue λ_i, $\mathbf{V}_s \in$

$\mathbb{R}^{p\times(p-q)}$, and $\mathbf{V}_n \in \mathbb{R}^{p\times q}$. The estimators for the stationary $\mathbf{W}_s$ and nonstationary $\mathbf{W}_n$ parts of the unmixing matrix from (2) are defined as $\widehat{\mathbf{W}}_s = \mathbf{V}_s^\top \mathbf{Q}_\mathbf{S}(\mathbf{x})^{-1/2}$ and $\widehat{\mathbf{W}}_n = \mathbf{V}_n^\top \mathbf{Q}_\mathbf{S}(\mathbf{x})^{-1/2}$, respectively.

The estimation method of Definition 2 is motivated by the sliced inverse regression (SIR), which is a supervised dimension reduction method [6]. In SIR one assumes to have a one-dimensional response y and a p-variate explanatory variable $\mathbf{x}$ such that $\mathbf{x}$ and y are conditionally independent given $\mathbf{U}\mathbf{x}$, where $\mathbf{U} \in \mathbb{R}^{k\times p}$ and $k < p$. That is, under the model, the dimension of the explanatory variable can be reduced from p to k dimensions. The matrix $\mathbf{U}$ spanning the lower dimensional subspace is estimated as follows. First, the whitened observations of the explanatory variable are divided into segments based on the values of the response variable. Then, the sample means corresponding to segments are used to compute a weighted covariance matrix similar to the matrix $\mathbf{M}$ of (1). Eigenvectors corresponding to the k largest eigenvalues of the weighted covariance matrix are used to acquire the estimate of $\mathbf{U}$. Thus, the method of Definition 2 is the SIR with a dummy response, which divides the spatiotemporal domain into segments.

We remark that in this work we restrict attention to non-stationarity in the mean. However, under the model specified in Definition 1, one could also consider cases where the component fields of $\mathbf{n}$ exhibit non-stationarity in the mean, variance, or autocorrelation. In such scenarios, as discussed in [4], one can compute different statistics—other than segment-wise means—and compare them to their global counterparts, formulating the task as a joint diagonalization problem. We will explore this extension in a follow-up paper.

4 Simulation Study

The finite sample performance of the stSSA method of Definition 2 is assessed with a simulation study. Throughout, we assume that the latent dimension q is known. For different choices of latent components and segmentation for the stSSA, we measure the method's performance in estimating the stationary $\mathbf{W}_s$ and nonstationary $\mathbf{W}_n$ parts of the unmixing matrix. However, as noted in Sect. 2, the unmixing matrix is not uniquely defined. Thus, the estimates $\widehat{\mathbf{W}}_s$ and $\widehat{\mathbf{W}}_n$ are not compared directly to $\mathbf{W}_s$ and $\mathbf{W}_n$, respectively. Instead, we compute the projection matrices $\mathbf{P}_s = \mathbf{W}_s^\top \left(\mathbf{W}_s\mathbf{W}_s^\top\right)^{-1} \mathbf{W}_s$ and $\mathbf{P}_n = \mathbf{W}_n^\top \left(\mathbf{W}_n\mathbf{W}_n^\top\right)^{-1} \mathbf{W}_n$ corresponding to the stationary and nonstationary subspaces. Similarly, let $\widehat{\mathbf{P}}_s$ and $\widehat{\mathbf{P}}_n$ be the projection matrices corresponding to $\widehat{\mathbf{W}}_s$ and $\widehat{\mathbf{W}}_n$. Then, as the performance measures for the stationary and nonstationary parts, we use the squared distances

$$\mathbf{s}_{\text{perf}} = \frac{1}{2}\left\|\mathbf{P}_s - \widehat{\mathbf{P}}_s\right\|_F^2 \quad \text{and} \quad \mathbf{n}_{\text{perf}} = \frac{1}{2}\left\|\mathbf{P}_n - \widehat{\mathbf{P}}_n\right\|_F^2, \tag{3}$$

where $\|\cdot\|_F$ is the Frobenius norm.

For a detailed treatment of the above performance measures, see [7]. The stSSA method is compared to a baseline method. In the baseline method, space-time segmentation is not used. Instead, we estimate the unmixing matrix with $\mathbf{W} = \mathbf{V}_{\text{rand}}\mathbf{Q}_{\mathbf{S}}(\mathbf{x})^{-1/2}$, where $\mathbf{V}_{\text{rand}}$ is sampled from the uniform distribution (the Haar measure) on orthogonal matrices. With a sufficiently large sample size and a sufficiently good choice of segmentation, the stSSA should outperform the baseline method with respect to performance measures given in Eq. (3).

As the domain for each simulation setting, we choose $\mathcal{U} \times \mathcal{T} = [0, 1]^2 \times \{1, \ldots, n_t\}$. Here, a *simulation setting* refers to a certain choice of latent components, and in total, we have seven different settings. In each setting, the ith component is of the form $z_i(\mathbf{u}, t) = \mu_i(\mathbf{u}, t) + \varepsilon_i(\mathbf{u}, t)$, where μ_i is a deterministic function giving the trend and ε_i is a zero-mean stationary field.

More precisely, the field ε_i is constructed by simulating two independent Gaussian fields $\varepsilon_i^{(1)} = \varepsilon_i^{(1)}(\mathbf{u})$ and $\varepsilon_i^{(2)} = \varepsilon_i^{(2)}(t)$. The covariance structure of the former field $\varepsilon_i^{(1)}$ is determined by the Matérn covariance function

$$C_{\nu,\phi}^{(1)}(\mathbf{u}_1, \mathbf{u}_2) = c_{\nu,\phi} \left(\frac{\|\mathbf{u}_2 - \mathbf{u}_1\|_2}{\phi} \right)^{\nu} K_\nu \left(\frac{\|\mathbf{u}_2 - \mathbf{u}_1\|_2}{\phi} \right),$$

where $c_{\nu,\phi}$ is a constant depending on ν and θ such that $\lim_{\mathbf{u}_1 \to \mathbf{u}_2} C_{\nu,\phi}^{(1)}(\mathbf{u}_1, \mathbf{u}_2) = 1$, K_ν is the modified Bessel function of the second kind, and $\| \cdot \|_2$ is the usual Euclidean norm. The covariance structure of the latter field $\varepsilon_i^{(2)}$ is determined by the exponential covariance function

$$C_\theta^{(2)}(t_1, t_2) = \exp\left(-\frac{|t_2 - t_1|}{\theta} \right).$$

Finally, the field ε_i is obtained by setting $\varepsilon_i(\mathbf{u}, t) = \varepsilon_i^{(1)}(\mathbf{u})\varepsilon_i^{(2)}(t)$.

For the trend function we consider the possibilities

$$\mu^{(1)}(\mathbf{u}, t) = \begin{cases} 0, & \text{if} \quad \mathbf{u} \in [0, 1] \times [0, 0.5] \\ 0.2, & \text{if} \quad \mathbf{u} \in [0, 1] \times (0.5, 1] \end{cases},$$

$$\mu^{(2)}(\mathbf{u}, t) = \begin{cases} -1, & \text{if} \quad t \le n_t/2 \\ -1.2, & \text{if} \quad t > n_t/2 \end{cases},$$

$$\mu^{(3)}(\mathbf{u}, t) = \begin{cases} 2, & \text{if} \quad \mathbf{u} \in [0, 0.5] \times [0, 1] \\ 2.2, & \text{if} \quad \mathbf{u} \in (0.5, 1] \times [0, 1] \end{cases},$$

$$\mu^{(5)}(\mathbf{u},t) = \begin{cases} 5, & \text{if } \mathbf{u} \in [0,0.5] \times [0,0.5] \times [1, n_t/2] \\ 5, & \text{if } \mathbf{u} \in (0.5,1] \times [0,0.5] \times [1, n_t/2] \\ -5, & \text{if } \mathbf{u} \in [0,0.5] \times (0.5,1] \times [1, n_t/2] \\ -5, & \text{if } \mathbf{u} \in (0.5,1] \times (0.5,1] \times [1, n_t/2] \\ 5, & \text{if } \mathbf{u} \in [0,0.5] \times [0,0.5] \times (n_t/2, n_t] \\ 5, & \text{if } \mathbf{u} \in (0.5,1] \times [0,0.5] \times (n_t/2, n_t] \\ -5, & \text{if } \mathbf{u} \in [0,0.5] \times (0.5,1] \times (n_t/2, n_t] \\ -5, & \text{if } \mathbf{u} \in (0.5,1] \times (0.5,1] \times (n_t/2, n_t] \end{cases}, \quad (4)$$

and lastly, $\mu^{(4)}(\mathbf{u},t)$ and $\mu^{(6)}(\mathbf{u},t)$ are segmented similarly in space and time as $\mu^{(5)}(\mathbf{u},t)$, but the values of the trend functions are different in the corresponding blocks. That is, the value of $\mu^{(4)}(\mathbf{u},t)$ decreases from -2 to -2.2 monotonically, and the value of $\mu^{(6)}(\mathbf{u},t)$ oscillates from 5 to -5 as one jumps from block to block from up to down in the order given by Eq. (4).

Now, we are ready to describe the latent components in the simulation settings in all of which we set $p = 2q$. For the stationary part we set $\mathbf{s} = (\varepsilon_1, \dots, \varepsilon_q)$, and for the nonstationary part we set $\mathbf{s} = (\mu^{(\pi(1))} + \varepsilon_{q+1}, \dots, \mu^{(\pi(q))} + \varepsilon_p)$ for some function $\pi : \{1, \dots, q\} \to \{1, \dots, 6\}$, where $\varepsilon_1 \dots \varepsilon_p$ are independent zero-mean Gaussian processes having dependence structure described by the parameters (ν, ϕ, θ). That is, for a setting we select a trend function for each nonstationary component, and the parameters giving the dependence structure (ν, ϕ, θ) are the same for each component.

Settings 1–6 are quite similar, but the dependence structure and the latent dimension p change from setting to setting. For settings i and $i+3$ we set $q = i+1, i \in \{1, \dots, 3\}$, and as the trend functions we choose $\mu^{(1)}, \dots, \mu^{(q)}$ for the nonstationary components. That is, the latent dimension is 4, 6, or 8 depending on the setting. Furthermore, for Settings 1–3 and 4–6 we set $(\nu, \phi, \theta) = (0.5, 10, 0.5)$ and $(\nu, \phi, \theta) = (0.5, 1, 2)$, respectively. In the case $(\nu, \phi, \theta) = (0.5, 10, 0.5)$ we have short-range dependence, and in the case $(\nu, \phi, \theta) = (0.5, 1, 2)$ we have long-range dependence. For Setting 7 we set $q = 2$, and for the nonstationary latent components we choose the trend functions $\mu^{(5)}$ and $\mu^{(6)}$. We also set $(\nu, \phi, \theta) = (0.5, 10, 0.5)$. That is, in Setting 7, both nonstationary components have an "oscillating" trend in the space-time blocks given by Eq. (4).

For each simulation setting, with a chosen sample size (n_s, n_t), simulation with either the baseline method or the stSSA with a chosen segmentation is repeated $m = 100$ times. In each round $j \in \{1, \dots, m\}$, the elements of $\mathbf{A}$ are generated from the uniform distribution on $[-1, 1]$. For Settings 1–6 we use the sample size $(n_s, n_t) = (50, 100)$, and for Setting 7 we use the sample size $(n_s, n_t) = (100, 300)$. For a setting the n_s spatial points are generated from a uniform distribution on $[0, 1]^2$; however, the spatial locations are fixed across the simulation rounds $j \in \{1, \dots, m\}$. That is, the value of the observed field $\mathbf{x}$ is recorded in time steps $\{1, \dots, n_t\}$ for fixed locations across simulation rounds. However, the mixing matrix is different for each simulation round. The estimates $\widehat{\mathbf{W}}_s$ and $\widehat{\mathbf{W}}_n$ for the stationary part and the nonstationary part of the unmixing matrix are acquired with the stSSA or the baseline method. Simulation round j is finished by computing the performance measures of Eq. (3) for the stationary and nonstationary parts.

In total, we have seven segmentation options for the stSSA, which are labeled S1–S7. To describe the segmentations we refer to the first spatial coordinate, the second spatial coordinate, and time by first, second, and third coordinates. In segmentations Si, $i \in \{1, 2, 3\}$, we do not segment the ith coordinate at all, but the two remaining coordinates are divided into two equally large segments. That is, in Segmentations S1–S3 one dimension is not segmented at all. In Segmentations S4–S7 all coordinates are divided into two, three, four, and ten equally large segments, respectively. For example, in Segmentation S5 the space-time is divided into 3^3 equally sized blocks. In Settings 1–6, all the segmentations are considered. However, in Setting 7 only Segmentations S1, S2, and S5 are used.

Table 1 shows the medians of the performance measures $\mathbf{s}_{\text{perf}}$ and $\mathbf{n}_{\text{perf}}$ computed from $m = 100$ simulation rounds for Settings 1–6 when the stSSA with Segmentations S1–S7 or alternatively, the baseline method (B) is used. For Settings 1–5 the median performances of both stationary and nonstationary parts corresponding to the best segmentation are bolded. For Settings 1–4 the best-performing segmentation corresponds to the true one. For example, in Setting 2 the trend is different in four blocks in space-time. Segmentation S4 matches the blocks in which the trends differ perfectly. Generally, Table 1 shows that as the dimension p increases and the short-range dependence $(\nu, \phi, \theta) = (0.5, 10, 0.5)$ is changed to the long-range dependence $(\nu, \phi, \theta) = (0.5, 1, 2)$, the performance of the stSSA deteriorates. This is most evident in Setting 6 where for any segmentation the stSSA performs almost equally poorly as the baseline method. Furthermore, in Setting 6 there is no unique segmentation for which the performance is the optimal for both stationary and nonstationary parts. Also, interestingly, for Setting 5, the best segmentation does not correspond to the theoretical one. This might be due to high dimension p and long-range dependence. Lastly, as the segmentation gets finer (from S4 to S7), the performance of the stSSA worsens due to the small sample size per segment $|\mathbf{S}_k|$.

Table 2 shows the simulation results for Setting 7. Similarly, as in Table 1, the median performance measures are shown for the stSSA with selected segmentations (S1, S2, and S5) and the baseline method. In addition, the average eigenvalues of the matrix $\mathbf{M}$ used in the stSSA for selected segmentations are given in the last four columns of the table. In this setting the trend changes only in space, not in time. Performance measures of Table 2 warn that it is not sufficient to choose any segmentation in space to acquire a sufficient performance. That is, because of the oscillating trend, if a wrong segmentation is chosen, the sample mean of some nonstationary latent component is close to zero in each segment. Eigenvalues corresponding to Segmentations S1 and S2 suggest that with both segmentations, the stSSA recognizes only one of the two nonstationary components.

Table 1 Simulation results for Settings 1–6. For each simulation setting, we use the stSSA with the chosen segmentation (S1–S7) or the baseline method (B). In each simulation round $j \in \{1, \ldots, m\}$, a random mixing matrix is used, but the sample size $(n_s, n_t) = (50, 100)$ is fixed. Each cell corresponds to the median computed from $m = 100$ performance measures for the estimator of the stationary or nonstationary part of the mixing matrix. For all settings, except for Setting 6, the medians corresponding to the best-performing segmentation are bolded. Abbreviations seg. and sub. stand for segmentation and subspace, respectively

		$(\nu, \phi, \theta) = (0.5, 10, 0.5)$			$(\nu, \phi, \theta) = (0.5, 1, 2)$		
Seg.	Sub.	$(q, p) = (2, 4)$	$(q, p) = (3, 6)$	$(q, p) = (4, 8)$	$(q, p) = (2, 4)$	$(q, p) = (3, 6)$	$(q, p) = (4, 8)$
S1	$\mathbf{s}_{\text{perf}}$	**0.10**	0.79	1.12	**0.40**	0.95	1.30
	$\mathbf{n}_{\text{perf}}$	**0.12**	0.65	1.11	**0.43**	0.90	1.40
S2	$\mathbf{s}_{\text{perf}}$	0.62	0.73	1.02	0.69	0.97	1.24
	$\mathbf{n}_{\text{perf}}$	0.62	0.63	1.10	0.66	0.93	1.26
S3	$\mathbf{s}_{\text{perf}}$	0.53	0.79	1.12	0.47	**0.83**	1.14
	$\mathbf{n}_{\text{perf}}$	0.59	0.77	1.17	0.68	**0.88**	1.26
S4	$\mathbf{s}_{\text{perf}}$	0.11	**0.41**	**0.92**	0.44	0.85	1.19
	$\mathbf{n}_{\text{perf}}$	0.12	**0.34**	**0.97**	0.49	0.91	1.21
S5	$\mathbf{s}_{\text{perf}}$	0.43	0.85	1.09	0.62	1.00	1.34
	$\mathbf{n}_{\text{perf}}$	0.49	0.86	1.11	0.72	0.89	1.32
S6	$\mathbf{s}_{\text{perf}}$	0.45	0.87	1.13	0.64	0.96	1.28
	$\mathbf{n}_{\text{perf}}$	0.43	0.80	1.21	0.66	0.95	1.38
S7	$\mathbf{s}_{\text{perf}}$	0.69	0.96	1.25	0.72	0.95	1.29
	$\mathbf{n}_{\text{perf}}$	0.75	0.96	1.24	0.76	0.91	1.28
B	$\mathbf{s}_{\text{perf}}$	0.66	0.98	1.32	0.65	1.00	1.30
	$\mathbf{n}_{\text{perf}}$	0.69	0.99	1.33	0.73	1.00	1.33
		Setting 1	Setting 2	Setting 3	Setting 4	Setting 5	Setting 6

Table 2 Simulation results for Setting 7. For each simulation setting, we use the stSSA with the chosen segmentation (S1, S2, or S5) or the baseline method (B). In each simulation round $j \in \{1, \dots, m\}$, a random mixing matrix is used, but the sample size $(n_s, n_t) = (100, 300)$ is fixed. Each cell in the first two columns corresponds to the median computed from $m = 100$ performance measures for the estimator of the stationary or nonstationary part of the mixing matrix. Cells of the remaining four columns give the average eigenvalues of the matrix $\mathbf{M}$ used in the stSSA computed from $m = 100$ simulation rounds. The average eigenvalues are ordered from the largest to the smallest $\lambda_1 \geq \cdots \geq \lambda_4$. Abbreviations seg. and sub. stand for segmentation and subspace, respectively

	Sub.		Eigenvalues			
Seg.	$\mathbf{s}_{\text{perf}}$	$\mathbf{n}_{\text{perf}}$	λ_1	λ_2	λ_3	λ_4
S1	0.64	0.80	9.6×10^{-1}	7.5×10^{-4}	3.9×10^{-5}	8.9×10^{-20}
S2	0.69	0.72	9.6×10^{-1}	8.1×10^{-4}	6.0×10^{-5}	5.1×10^{-18}
S5	1.0×10^{-5}	1.5×10^{-3}	6.8×10^{-1}	5.6×10^{-1}	5.5×10^{-3}	1.6×10^{-3}
B	0.083	0.95	–	–	–	–

For reproducibility, the simulation codes are provided in https://github.com/perej1/sim-temporal-spatial-ssa/releases/tag/v1.1. Also, related to simulation scenarios of Tables 1 and 2, boxplots illustrating the scatter of the performance measures are provided in the GitHub repository.

5 Conclusions

We developed a statistical method for separating the stationary and nonstationary latent subspaces spanned by the matrices $\mathbf{A_s}$ and $\mathbf{A_n}$ from Definition 1, respectively. The practicality of the method is illustrated with a simulation study. The simulation study suggests that, for a fixed sample size, the performance of the method suffers in the presence of long-range dependence or if the latent dimension is high. Furthermore, Simulation Setting 7 indicates that, in some cases, the segmentation has to be chosen carefully to acquire satisfactory results. Thus, it is advisable to perform the method with multiple segmentations. Nonetheless, if the segmentation is too fine, the method might fail due to an insufficient sample size per segment.

Throughout the simulations, we assume that the dimension of the nonstationary subspace q is known. However, in practical settings, one usually does not know the value of q. Thus, methods for estimating q are crucial. As Table 2 shows, for a sufficiently large sample size and a reasonable segmentation, exactly the q largest eigenvalues of the matrix $\mathbf{M}$ from (1) differ significantly from zero. Thus, the magnitudes of the eigenvalues of $\mathbf{M}$, or the scree plot, can be used to guess the dimension q. Furthermore, in the context of spatiotemporal data, detecting combined non-stationarity in mean, variance, and covariance is a topic of future research. Also under additional assumptions the actual latent components might be fully identifiable following considerations of [8, 11].

Acknowledgments The position of JP is funded by the Research Council of Finland (363261). We also wish to acknowledge CSC—IT Center for Science, Finland, for computational resources.

Bibliography

1. Berild, M.O., Fuglstad, G.A.: Non-stationary spatio-temporal modeling using the stochastic advection–diffusion equation. Spat. Stat. **64**, 100867 (2024)
2. Chen, W., Genton, M.G., Sun, Y.: Space-time covariance structures and models. Ann. Rev. Stat. Its Appl. **8**, 191–215 (2021)
3. De Iaco, S., Cappello, C., Palma, M., Nordhausen, K.: A multivariate approach for modeling spatio-temporal agrometeorological variables. Environmetrics **36**(2), e2891 (2025)
4. Flumian, L., Matilainen, M., Nordhausen, K., Taskinen, S.: Stationary subspace analysis based on second-order statistics. J. Comput. Appl. Math. **436**, 115379 (2024)
5. Kyriakidis, P.C., Journel, A.G.: Geostatistical space–time models: a review. Math. Geol. **31**, 651–684 (1999)
6. Li, K.C.: Sliced inverse regression for dimension reduction. J. Am. Stat. Assoc. **86**, 316–327 (1991)
7. Liski, E., Nordhausen, K., Oja, H., Ruiz-Gazen, A.: Combining linear dimension reduction subspaces. In: Agostinelli, C., Basu, A., Filzmoser, P., Mukherjee, D. (eds.) Recent Advances in Robust Statistics: Theory and Applications, pp. 131–149. Springer India, New Delhi (2016)
8. Muehlmann, C., De Iaco, S., Nordhausen, K.: Blind recovery of sources for multivariate space-time random fields. Stoch. Environ. Res. Risk Assess. **37**, 1593–1613 (2023)
9. Porcu, E., Furrer, R., Nychka, D.: 30 years of space-time covariance functions. Wiley Interdisciplinary Reviews. Comput. Stat. **13**(2), e1512 (2021)
10. Shand, L., Li, B.: Modeling nonstationarity in space and time. Biometrics **73**(3), 759–768 (2017)
11. Sipilä, M., Cappello, C., De Iaco, S., Nordhausen, K., Taskinen, S.: Modelling multivariate spatio-temporal data with identifiable variational autoencoders. Neural Netw. **181**, 106774 (2025)
12. Wu, X., Zhan, H., Hu, J., Wang, Y.: Non-stationary GNNCrossformer: transformer with graph information for non-stationary multivariate spatio-temporal wind power data forecasting. Appl. Energy **377**, 124492 (2025)

Part II
Environmental Control and Integration

Space Turns to Time: The Advent of Time Series of Remote Sensing Images

Alfred Stein

Abstract Spatial statistics is traditionally concerned with data collected in space. As one field of application that has developed in the recent past, we consider satellite images. Starting with using such images as collateral information for supporting the handling and interpretation of spatial data, we later explored the images themselves, like for predicting pixel values below the clouds. The issue of up- and downscaling was addressed, as well as the integration of images, which was followed by including images into the modeling of spatial phenomena. At present, there is more and more attention to time series of images, with new challenges and opportunities emerging. Both optical and radar images require a solid understanding of the quality of the data and a fine-tuning to properly relate successive images. In this presentation, I will present some of the recent developments, like those developed in Zhang et al. [9], Mohammadi et al. [4], and Kulshrestha et al. [3]. Attention is given to issues of deep learning in the context of modern AI.

Keywords Spatial statistics · Air quality · Satellite images · Time series · Integration

1 Air Quality

Air quality has been a topic of large research interest in the previous decades. Research, in particular, concerned aerosols, NOx, SO_2, and other substances. Traditionally, data are collected on the ground by a variety of sensors, from wearables through low-cost sensors to highly dedicated representative measurement stations. In the Netherlands, there are some 25 of the latter, within the city of Eindhoven a measurement network of 35 low-cost sensors was located, while

A. Stein (✉)
Faculty of Geo-Information Science and Earth Observation (ITC), University of Twente, Enschede, The Netherlands
e-mail: a.stein@utwente.nl

S. De Iaco et al. (eds.), *Exploration of Spatio-Temporal Environmental Conditions: Harmonized Databases and Analytical Techniques*, Springer Proceedings in Mathematics & Statistics 531, https://doi.org/10.1007/978-3-032-17526-7_5

wearables are widely available. A low-cost urban air quality sensor network has been used to study the spatiotemporal variability in air pollutant concentrations. It was set up in 2013 as part of the civil initiative AiREAS. Research has been done on evaluating the data quality of the collected data and its usability in spatiotemporal modeling and health effect estimation. The data are collected at 35 locations, on the variables like PM_{10}, $PM_{2.5}$, PM_1, and NOx. The temporal resolution is 10 minutes. Let me start with giving two motivating studies from the recent past that we addressed on this network.

1.1 Outliers

Specific interest has focused on outliers in relation to health effects and spreading models. Outliers in the measurements could reflect measurement errors or unusually high or low air pollution events. Van Zoest et al. [7] presented an outlier detection method based upon a spatiotemporal classification. The focus was on hourly nitrogen dioxide (NO_2) concentrations, as NO_2 has a large spatiotemporal variability and has a strong association with health effects, and the study site was the city of Eindhoven, the Netherlands, where a pioneering monitoring network was set up. They defined different spatiotemporal classes that reflect urban background vs. urban traffic stations, weekdays vs. weekends, and four periods per day and used truncated normal distributions to set thresholds to identify outliers in each spatiotemporal class. Based upon this study, they presented a method that is able to detect outliers in recognizing the large spatiotemporal variability of urban air pollutant concentrations.

1.2 Spatiotemporal Interpolation

Of importance is the relatively scarcity of all monitoring networks: The number of stations is necessarily limited, while the region to be represented may large, and the spatial and temporal variability of several substances is likely to be high. Hence, air pollutant concentrations have to be predicted at unobserved locations and times. Van Zoest et al. [8] used spatiotemporal regression kriging to map NO_2 at a 25 m spatial resolution and at hourly temporal resolution. The trend was modeled separately from autocorrelation in the residuals and consists of population density and road type as spatial variables and meteorological variables as temporal covariates including. Spatiotemporal autocorrelation was modeled by fitting a sum-metric spatiotemporal variogram model. They showed that this method provides local estimates of the strength and association of air pollution sources and sinks and allows for near real-time prediction of air pollutant concentrations. The resulting maps visualized these concentrations in space and time and can be used to assess exposure for the evaluation of short-term health effects.

1.3 *Remote Sensing Images*

Currently, more and more satellites collect air quality indicators. Research on remote sensing offers a fascinating world, where the amount of data is incredible and is increasing. These images are big data, of different types of pixels: optical and radar, with a few to hundreds of spectral bands. Initially, in spatial statistical studies, the main interest was to use such data as a covariable in the statistical analyses. The next stage was to do the analysis on the images themselves, like the prediction of pixels under the clouds. Several studies expanded this research to move from pixels to objects, i.e., a classification first to generate the objects of interest, like studies to coastal inundations and Antarctic ice objects. Of importance as well is the challenge of up- and downscaling, from combining pixels or objects towards trying to identify objects smaller than the pixel size, the so-called super-resolution mapping. In addition, the constellation of satellites is changing almost daily. A big step forward came with the free availability of Landsat and Sentinel images. A particular satellite that is of interest for air quality monitoring concerns the Sentinel 5 satellite, which is able to collect several air quality substances on the ground surface. The satellite has a return time of a single day and measures the instantaneous quality averaged over areas of 5 by 5 km.

2 Time Series of Images

Time series of images are offering new opportunities and require dedicated methods to obtain the maximum benefit. In recent research, attention has been given on developing methods for integrating these images. As the Earth surface is changing, images can be used to monitor those changes, and thus they require methods from both time series and spatial statistics or their combination to extract the relevant information. In many instances, the interest centers on tracking and following objects. Such objects should be well-defined up-front but may have a functional, a fuzzy, or an irregular shape. The challenge is then to follow the objects.

A particular example that we studied in the past was the tracking of bush fires [6]. In that study we used the Gaussian function for representing the object forest fire, while a kernel convolution method to fit the data over space. Object identification concerned the fire object by its location and the uncertainties. To relate a fire object at one moment to the same object at the successive time, we related two objects with the lowest distance at adjacent time images, thus ensuring connectivity in time. We initially extracted n fires from each image, where n is the number of possible fires.

The analysis considered both the identification and the changes that may occur when handling objects. Thus, object identification in relation to the tracking was analyzed for:

- Merging, corresponding to reducing the number of fire objects with 1. This was handled by reducing n with 1 and averaging the centroids.
- Splitting, i.e., increasing the number of fire objects with 1. Splitting required an initial starting point of two fires out of a single one and interactive visual methods were applied for this purpose.
- Detecting anomalies. For instance, such an anomaly is an identified object of a sufficiently high temperature but with a too large distance to fires at the previous or the succeeding times to be realistic as a fire object.

This process was automated using an ordinary least square fitting. Once fitted, the function is subtracted from the main image leading to a residual image $\epsilon(s)$ and a function $\phi(s)$, both depending upon the location s. It is summarized as

$$\epsilon(s) = I(s) - h \cdot \phi(s) - m, \tag{1}$$

where h is a scaling parameter, and m is the global minimum of $I(s)$, while for $\phi(s)$ we used the Gaussian bivariate function, i.e.,

$$\phi(s) = \frac{1}{2\pi \cdot \sigma} \cdot exp\left[-\frac{1}{2} \cdot \left(\frac{s-\mu}{\sigma} \right)^2 \right]. \tag{2}$$

Here, μ is the center location of the fire object, and σ measures the extent of the fire along two axes. This process is repeated k times, until no further maximum is identified that could signify a fire.

The final result equals

$$\epsilon_{res}^{(k)}(s) = I(s) - h^{(1)}\phi^{(1)}(s) - \ldots - h^{(k)}\phi^{(k)}(s) - m, \tag{3}$$

where the $\{\phi^{(j)}(s)\}_{j=1,\ldots,k}$ equal the k fitted functions with parameters $\sigma^{(j)}$ and $\mu^{(j)}$, the $\{h^{(j)}\}_{j=1\ldots\, k}$ are the k scaling parameters, and $m = m^{(1)} + \ldots + m^{(k)}$ is the sum of the minimum values after each iteration. Each function $\phi^{(j)}(s)$ represents a fire object. The choice for the value of n is set heuristically equal to ten, as an image of the size of 54 km^2 unlikely covers more than ten fires, as identifiable from SEVIRI and as being relevant to fire fighting.

To track these objects, we need to include the temporal dimension as well, resulting in a functional relation that is defined as

$$\hat{s}_{t+4} = f(s_t, \sigma_t, WD_t, WS_t, NDVI_t), \tag{4}$$

where $\hat{s}_{t+4}$ is the predicted location of the object at 4 units ahead of t, s_t the centroid location of the fire at time t, σ_t its spread, WD_t, WS_t, and $NDVI_t$ the direction of the wind, the speed of the wind, and the NDVI at that time, respectively. As the NDVI is typically obtained from remote sensing images, we notice here the essential use of the Meteosat data, in connection with the field data. Observations were available every 15 minutes, while predictions were considered relevant in 1 hour of time from t, and hence the subscript on the left side is set equal to $t+4$. Tracking was used to study the change of objects in time and the effects of vegetation and the movement of fire, to predict where the fire would be in 1 hour. Space is turning to time.

3 Integrating the Objects

Using modern deep learning methods, we are increasingly able to give more attention to time series of images. Several models have been developed. In [3] we classified forms of anomalous behavior with a deep learning-based supervised time series classification. Here we used a two-layered bidirectional long short term memory (LSTM) classification model. In [4] we used a 3D Fully Convolutional Network (FCN), which was successfully investigated in crop type mapping by some previous studies. 3D-convolution operators allow the 3D FCN to learn both spatial and temporal relationships in image time series. To address the challenge of visualization, we used the t-distributed stochastic neighbor embedding method $t-SNE$. It is a statistical method for visualizing high-dimensional data by giving each data point a location in a two- or three-dimensional representation. This nonlinear dimensionality reduction technique reduces the original high-dimensional data into two (or three) dimensions, in particular for the purpose of visualization. In [4] we were able to apply the 3D FCN trained using three samples on images of 128×128 pixels covering 23 time steps in the time series of images and six spectral bands. This 3D FCN was used to transfer from one site towards other sites with similar characteristics. In particular, it was possible with reasonably good results to train a deep learning network on one site during a range of years and tested on another site during a successive year.

4 Ongoing Research

In ongoing research, time series of images are more and more prominent. We note that the type of images is playing a key role. Optical images like the freely available Landsat images are useful to link for processes that are well captured at spatial resolutions of 30 m and temporal resolutions of 16 days, for instance, urban expansion and coastal developments are well captured. In a study on the Wadden sea area [9], we explored the use 199 Sentinel-1 GRD, 132 Radarsat-2 SLC, and 157 Landsat images, to analyze coastal dynamics in the Dutch Wadden Sea from 1986 to 2020. Our Digital Elevation Model (DEMs) were validated against LiDAR data (2016–2019) and 58 ground anchor measuring stations (2011–2020). We again used an Object-Based Image Segmentation (OBIS) method, specifically designed for SAR images, to extract waterlines and distinguish tidal flats and shorelines from water bodies. This method integrates SAR polarimetric feature analysis to select high-quality images and employ partition processing to preserve local feature statistics. The proposed OBIS-based framework demonstrated its effectiveness for mapping national-scale tidal flats and sandbanks dynamics. These and other studies show that the future will be in the integration of multiple data sources, such as different types of satellite images and point data.

5 Looking Forward

Spatiotemporal data represent information that includes both spatial (location) and temporal (time) attributes, allowing for the analysis of how phenomena change over space and time. This type of data is crucial for understanding complex systems in various fields like urban planning, climate science, and transportation.

Important developments on which already much research has been done consider the modeling of spatiotemporal covariance functions. Such functions are able to describe the spatial and temporal dependencies of a response variable. For spatiotemporal data, the resolutions in space and time are critical. These units should well describe the expected resolution of the process behind the data.

For several studies, the use of an anisotropy parameter to relate space and time is adequate [5]. Such an analysis, however, lacks causality, and causal statistical inference is increasingly considered an important direction of research [1]. The main directions are finding cause-effect relationships behind observed phenomena. Much progress in causal inference has been made already, in studies on agriculture and the environment, while recent progress is offered in spatial statistics, where the main challenge to deal with was the nonrandom, non-repeatability, and synchronism of spatial data. The approach taken was to first refine the issues of causal inference and then discuss causal inference in spatial statistics.

A problem recently solved was that causal inference has been largely based on the use of some advanced temporal causation models, while temporal models have serious limitations when time series data are not available or present in significant variations. This causes a common challenge, for instance, in Earth system applications. In [2] spatial causation models are proposed that more completely explore spatial cross-sectional data in Earth systems. Using the generalized embedding theorem observations can be integrated to construct the state space of the dynamic system, while if two variables are from the same dynamic system, they are causally linked. The study shows that a Geographical Convergent Cross Mapping (GCCM) model is suited for spatial causal inference with spatial cross-sectional data-based cross-mapping prediction in reconstructed state space. GCCM is able to detect weak-moderate causation even if the correlation is not significant. While if the coupling between two variables is significant and strong, GCCM is advantageous in identifying the primary causation direction and better revealing the bidirectional asymmetric causation.

6 Concluding Remarks

Time series of remote sensing images offer an interesting opportunity to study Earth surface problems and to monitor its changes. A particular research gap remains the wide availability of sound methodologies that are able to make the most out of these time series. Several issues have been solved, but there are still important research gaps to be addressed. The frequency, areal resolution, and characteristics of the time series do not allow addressing all relevant questions, but many questions can now be answered in a coherent way, while new satellites emerging in the future will be helpful to further advance the opportunities. In the domain of spatial statistics, one major requirement is to face the challenge of including the spatiotemporal covariance structure within the large volumes of data offered by the time series.

Bibliography

1. Gao, B., Wang, J., Stein, A., Zeng, Z.: Causal inference in spatial statistics. Spat. Stat. **50**, 100621 (2022)
2. Gao, B., Yang, J., Zheng, C., Sugihara, G., Li, M., Stein, A., Kwan, M.P., Wang, J.: Causal inference from cross-sectional earth system data with geographical convergent cross mapping. Nat. Commun. **14**, 5875 (2023)
3. Kulshrestha, A., Chang, L., Stein, A.: Use of LSTM for sinkhole-related anomaly detection and classification of InSAR deformation time series. IEEE J. Sel. Top. Appl. Earth Obs. Remote Sens. **15**, 4559–4570 (2022)
4. Mohammadi, S., Belgiu, M., Stein, A.: Improvement in crop mapping from satellite image time series by effectively supervising deep neural networks. ISPRS J. Photogramm. Remote Sens. **198**, 272–283 (2023)
5. Stein, A., Van Groenigen, J.W., Jeger, M.J., Hoosbeek, M.R.: Space-time statistics for environmental and agricultural related phenomena. Environ. Ecol. Stat. **5**, 155–172 (1998)
6. Umamaheshwaran, R., Bijker, W., Stein, A.: Image mining for modeling of forest fires from meteosat images. IEEE Trans. Geosci. Remote Sens. **45**(1), 246–253 (2007)
7. van Zoest, V.M., Stein, A., Hoek, G.: Outlier detection in urban air quality sensor networks. Water Air Soil Poll. **229**, 111 (2018)
8. van Zoest, V., Osei, F.B., Hoek, G., Stein, A.: Spatio-temporal regression kriging for modelling urban NO_2 concentrations. Int. J. Geograph. Inf. Sci. **34**(5), 851–865 (2019)
9. Zhang, B., Chang, L., Wang, Z., Wang, L., Ye, Q., Stein, A.: Multi-decadal Dutch coastal dynamic mapping with multi-source remote sensing imagery. Int. J. Appl. Earth Sci. Geoinf. **138**, 104452 (2025)

Ensemble Smoother with Multiple Data Assimilation for Atmospheric Dust Source Identification: A Generic Framework Approach

J. Jaime Gómez-Hernández, Valeria Todaro, Marco D'Oria, and Maria Giovanna Tanda

Abstract The identification of atmospheric dust sources represents a critical challenge in environmental monitoring and climate research. The potential application of the Ensemble Smoother with Multiple Data Assimilation (ES-MDA) methodology, implemented through the generic open-source software package genES-MDA, to address the inverse problem of dust source identification in atmospheric systems is presented. Building upon successful applications in groundwater contaminant source identification and hydrological inverse modeling, we discuss how the same theoretical framework and computational tools can be adapted to atmospheric dust source identification by appropriately modifying the state equations and numerical codes. The genES-MDA package provides a model-independent framework that requires only access to a Python interface or command-line interface to the forward model, making it suitable for integration with atmospheric transport models. This approach offers a systematic methodology for uncertainty quantification and parameter estimation in atmospheric dust source identification problems. It extends the proven capabilities of ensemble-based data assimilation methods to atmospheric environmental applications, maintaining their fundamental advantages while addressing the specific characteristics of atmospheric transport processes.

Keywords Ensemble smoother · Multiple data assimilation · Atmospheric dust · Source identification · Inverse modeling · genES-MDA

J. J. Gómez-Hernández (✉)
Institute for Water and Environmental Engineering, Universitat Politècnica de València, Valencia, Spain
e-mail: jaime@dihma.upv.es

V. Todaro · M. D'Oria · M. G. Tanda
Department of Engineering and Architecture, Università di Parma, Parma, Italy

S. De Iaco et al. (eds.), *Exploration of Spatio-Temporal Environmental Conditions: Harmonized Databases and Analytical Techniques*, Springer Proceedings in Mathematics & Statistics 531, https://doi.org/10.1007/978-3-032-17526-7_6

1 Introduction

The identification and quantification of atmospheric dust sources constitute fundamental challenges in atmospheric science and environmental monitoring. Dust emissions have a significant impact on air quality, climate dynamics, and human health, making accurate source identification essential for effective environmental management and policy development. Traditional approaches to dust source identification often rely on forward modeling combined with observational data, but the inverse problem of determining source characteristics from downstream observations remains computationally challenging and methodologically complex.

Inverse problems in environmental sciences are inherently ill-posed, requiring sophisticated mathematical and computational approaches to obtain meaningful solutions. The Ensemble Smoother with Multiple Data Assimilation (ES-MDA) has emerged as a powerful methodology to address such inverse problems, particularly in subsurface hydrology and geophysical applications [2]. The success of ES-MDA in these domains suggests its potential applicability to the identification of atmospheric dust sources, provided that appropriate adaptations are made to account for the specific characteristics of atmospheric transport processes.

Current approaches to atmospheric dust source identification rely primarily on satellite-based techniques using aerosol optical depth, brightness temperature differences, or plume motion detection [4, 7, 13]. However, these methods face limitations including saturation in optically thick dust plumes, difficulties in distinguishing between dust sources and transport regions, and sensitivity to meteorological conditions. The development of inverse modeling approaches for dust source identification represents a complementary methodology that can address some of these limitations.

The genES-MDA software package represents a generic, open-source implementation of the ES-MDA methodology that has been successfully applied to various inverse modeling problems [10]. Originally developed for hydrogeological applications, the model-independent design of the package makes it suitable for extension to atmospheric applications. The software has demonstrated effectiveness in the identification of the source of contaminants in aquifers, the estimation of parameters in groundwater systems, and reverse flow routing problems [9], establishing a foundation for its application to the identification of atmospheric dust sources.

This work explores the theoretical basis and practical implementation considerations for applying the genES-MDA framework to atmospheric dust source identification. We discuss the necessary modifications to the state equations and numerical implementations required to adapt the methodology from subsurface to atmospheric applications. The approach maintains the fundamental advantages of ensemble-based methods, including uncertainty quantification and the ability to handle nonlinear, high-dimensional parameter spaces characteristic of atmospheric transport problems.

2 Theoretical Background

2.1 Ensemble Smoother with Multiple Data Assimilation

The Ensemble Smoother with Multiple Data Assimilation (ES-MDA) extends the classical Ensemble Smoother (ES) by assimilating the same dataset multiple times through iterative updates [11]. This approach addresses the limitations of single-step ensemble smoothers when dealing with highly nonlinear problems, such as those commonly encountered in atmospheric transport modeling.

The fundamental equation governing the ES-MDA update process can be expressed as

$$\mathbf{x}_j^{i+1} = \mathbf{x}_j^i + \mathbf{C}_{xy}^i(\mathbf{C}_{yy}^i + \alpha_i \mathbf{C}_D)^{-1}(\mathbf{d}_j^i - \mathbf{y}_j^i) \quad (1)$$

where $\mathbf{x}_j^{i+1}$ represents the updated parameter vector for the ensemble member j at iteration $i + 1$, $\mathbf{d}_j^i$ represents perturbed observations, and $\mathbf{y}_j^i$ represents the model predictions at the observation locations, obtained by running the forward model using $\mathbf{x}_j^i$ as input. $\mathbf{C}_{xy}^i$ is the cross-covariance matrix between parameters and model predictions, $\mathbf{C}_{yy}^i$ is the autocovariance matrix of model predictions, α_i is the inflation factor for iteration i, and $\mathbf{C}_D$ is the observation error covariance matrix.

The iterative nature of ES-MDA allows for gradual convergence towards the optimal solution while maintaining ensemble spread and providing uncertainty quantification. The inflation factors α_i must satisfy the constraint $\sum_i \alpha_i^{-1} = 1$, with i being the iteration number, to ensure proper statistical properties of the final ensemble.

2.2 Adaptation to Atmospheric Systems

The application of ES-MDA to atmospheric dust source identification requires careful consideration of the specific characteristics of atmospheric transport processes. Unlike groundwater systems, atmospheric transport involves complex three-dimensional flow patterns, turbulent mixing, and time-varying meteorological conditions that significantly influence dust dispersion patterns.

The forward model for atmospheric dust transport typically involves solving the advection-diffusion equation with appropriate source terms [1, 8]:

$$\frac{\partial C}{\partial t} + \nabla \cdot ((\mathbf{u} + \mathbf{v}_s)C) = \nabla \cdot (\mathbf{K}\nabla C) + S(\mathbf{x}, t) \quad (2)$$

where C represents dust concentration, $\mathbf{u}$ is the wind velocity field, $\mathbf{v}_s$ is the settling velocity, $\mathbf{K}$ is the turbulent diffusion tensor, and $S(\mathbf{x}, t)$ represents the source term to be identified through the inverse modeling process. This equation is formally the same as the advection-dispersion equation used to model mass transport in the subsurface.

The state vector in the atmospheric application includes parameters such as source location coordinates, emission rates, temporal emission patterns, and potentially meteorological parameters if they are subject to uncertainty. The observation vector consists of dust concentration measurements at various spatial and temporal locations, which may include ground-based monitoring stations, satellite observations, or aircraft measurements.

2.3 genES-MDA Framework Architecture

The genES-MDA package provides a flexible framework for implementing ES-MDA across different application domains. The software architecture separates ensemble management, data assimilation algorithms, and forward model interfaces, allowing for straightforward adaptation to new problem types.

Key components of the genES-MDA framework include parameter management for handling spatial and temporal parameter distributions, forward model interfaces providing standardized communication with external models through Python APIs or command-line interfaces, data assimilation engines implementing core ES-MDA algorithms with options for covariance localization and inflation, uncertainty quantification capabilities generating ensemble statistics and uncertainty bounds, and analysis tools for result interpretation.

The modular design of genES-MDA facilitates its extension to atmospheric applications by allowing users to implement custom forward model interfaces while leveraging the existing data assimilation infrastructure.

3 Methodology for Atmospheric Dust Source Identification

3.1 Problem Formulation

The atmospheric dust source identification problem can be formulated as an inverse modeling problem where the objective is to estimate source characteristics based on observed dust concentrations at monitoring locations. The problem involves determining the spatial distribution, temporal evolution, and intensity of dust emissions that best explain the observed concentration patterns.

The parameter vector typically includes source location coordinates, emission rates as functions of time, source geometry parameters, and potentially uncertain meteorological parameters. The observation vector consists of dust concentration measurements at various spatial locations and time instances. The forward model relates the source parameters to the predicted concentrations through the solution of the atmospheric transport equation.

3.2 Forward Model Requirements

The successful application of genES-MDA to atmospheric dust source identification requires a forward model capable of simulating dust transport and dispersion in the atmosphere. The forward model must satisfy several requirements: computational efficiency to allow for ensemble simulations with hundreds or thousands of realizations, interface compatibility providing either Python or command-line interfaces compatible with genES-MDA requirements, flexibility on parameter handling allowing modification of source parameters and meteorological inputs, and appropriate output format generating concentration predictions at observation locations in genES-MDA-compatible data structures.

Several atmospheric transport models could potentially serve as forward models, including Lagrangian particle dispersion models, Eulerian grid-based models, and simplified analytical solutions for specific scenarios.

3.3 Implementation Considerations

The adaptation of genES-MDA to atmospheric dust source identification involves several implementation considerations. Appropriate spatial and temporal discretization of the parameter space is essential to accurately represent source characteristics while maintaining computational feasibility. This may involve defining source regions rather than point sources and discretizing emission rates over time intervals.

Observation processing requires handling dust concentration observations to account for measurement uncertainties, detection limits, and temporal averaging. The observation error covariance matrix must reflect these uncertainties appropriately. Ensemble generation must represent prior knowledge and uncertainty about source characteristics, potentially involving topographic information, land use data, and historical emission patterns.

Given the spatial extent of atmospheric transport problems, covariance localization techniques may be necessary to prevent spurious correlations between distant locations.

3.4 Comparison with Groundwater Applications

The extension of genES-MDA from groundwater to atmospheric applications involves several key differences and similarities. Both domains involve transport processes governed by advection-diffusion equations, require identification of source characteristics from downstream observations, benefit from ensemble-based uncertainty quantification, and involve high-dimensional parameter spaces.

However, atmospheric transport involves three-dimensional flow fields compared to typically two-dimensional groundwater flow, operates on different time scales (hours to days for atmospheric versus years to decades for groundwater), includes meteorological variability introducing additional complexity, and utilizes different observation networks and measurement techniques.

These differences require careful adaptation of the methodology while maintaining the fundamental advantages of the ES-MDA approach.

4 Applications and Advantages

4.1 Potential Applications

The methodology can be applied to various dust source identification problems. Desert dust sources identification in major desert regions supports climate modeling and air quality forecasting efforts [3, 5]. Industrial emissions quantification from mining operations, construction sites, and industrial facilities aids regulatory compliance and environmental impact assessment [6, 12]. Agricultural dust emission sources assessment from tillage operations and livestock facilities. Dust emission identification following natural disasters, such as volcanic eruptions or large-scale fires, supports rapid response and risk evaluation.

Each application presents unique challenges in terms of source characteristics, observation availability, and temporal scales, requiring careful adaptation of the methodology.

4.2 Advantages of the ES-MDA Approach

The application of ES-MDA to atmospheric dust source identification offers several significant advantages. The ensemble-based approach provides natural uncertainty quantification for estimated source parameters, enabling risk-based decision-making and confidence assessment. ES-MDA can handle the nonlinear relationships between source parameters and observations characteristic of atmospheric transport problems.

The methodology can simultaneously assimilate multiple types of observations, including concentration measurements, meteorological data, and auxiliary information. Compared to other inverse modeling approaches, ES-MDA offers good computational efficiency while maintaining statistical rigor. The genES-

MDA framework can work with various atmospheric transport models, providing flexibility in model selection and allowing for model comparison studies.

4.3 Limitations and Challenges

Despite its advantages, the ES-MDA approach faces several limitations. The methodology requires accurate and computationally efficient forward models, which may not be available for all atmospheric transport scenarios. The quality of source identification depends heavily on the spatial and temporal coverage of the observation network, which may be limited in many regions.

Uncertainties in meteorological conditions can significantly impact the accuracy of source identification, particularly for long-range transport scenarios. While efficient compared to some alternatives, the ensemble-based approach still requires significant computational resources for large-scale applications. Some source parameters may not be identifiable from available observations, particularly when multiple sources are present or when transport distances are large.

5 Future Developments

Several areas of development could enhance the application of ES-MDA to atmospheric dust source identification. Advanced localization techniques could account for the anisotropic nature of atmospheric transport and varying correlation scales in different meteorological conditions. Multi-scale approaches could integrate multiple spatial and temporal scales to address the range of processes involved in dust emission and transport.

Hybrid methods combining ES-MDA with other inverse modeling approaches could leverage the strengths of different methodologies. Real-time implementation could provide near-real-time source identification capabilities for emergency response and operational applications. The genES-MDA software package could be enhanced with atmospheric model interfaces, improved visualization tools, performance optimizations for larger spatial domains and finer temporal resolution, and cloud computing integration for large-scale operational applications.

6 Conclusions

This work has presented the theoretical foundation and practical considerations for applying the Ensemble Smoother with Multiple Data Assimilation methodology to atmospheric dust source identification. The genES-MDA software package provides a robust and flexible framework that can be adapted from its successful applications in groundwater systems to atmospheric environmental problems.

The key advantages of the ES-MDA approach include its ability to handle nonlinear inverse problems, provide uncertainty quantification, and integrate multiple types of observational data. The model-independent design of genES-MDA facilitates its application to various atmospheric transport models, making it a versatile tool for dust source identification across different scales and applications. The adaptation from groundwater to atmospheric applications requires careful consideration of the differences in transport processes, time scales, and observation characteristics. However, the fundamental mathematical framework remains applicable, and the proven success of ES-MDA in related domains provides confidence in its potential for atmospheric applications.

Future work should focus on developing specific implementations for atmospheric dust source identification, including integration with atmospheric transport models, validation using synthetic and real-world test cases, and development of operational capabilities for environmental monitoring and management applications.

The methodology presented here represents a step toward more systematic and quantitative approaches to atmospheric dust source identification, with potential benefits for air quality management, climate research, and environmental policy development. The open-source nature of the genES-MDA package ensures that these capabilities will be accessible to the broader research and operational communities.

Bibliography

1. Arya, S.P.: Air Pollution Meteorology and Dispersion. Oxford University Press, Oxford (1999)
2. Emerick, A.A., Reynolds, A.C.: Ensemble smoother with multiple data assimilation. Comput. Geosci. **55**, 3–15 (2013)
3. Evan, A.T., Flamant, C., Fiedler, S., Doherty, O.: An analysis of aeolian dust in climate models. Geophys. Res. Lett. **41**(16), 5996–6001 (2014)
4. Ginoux, P., Prospero, J.M., Gill, T.E., Hsu, N.C., Zhao, M.: Global-scale attribution of anthropogenic and natural dust sources and their emission rates based on MODIS deep blue aerosol products. Rev. Geophys. **50**(3), 1–36 (2012)
5. Kok, J.F., Adebiyi, A.A., Albani, S., Balkanski, Y., Checa-Garcia, R., Chin, M., et al.: Improved representation of the global dust cycle using observational constraints on dust properties and abundance. Atmos. Chem. Phys. **21**(10), 8127–8167 (2021)
6. Patra, A.K., Gautam, S., Kumar, P.: Emissions and human health impact of particulate matter from surface mining operation—A review. Environ. Technol. Innov. **5**, 233–249 (2016)
7. Schepanski, K., Tegen, I., Laurent, B., Heinold, B., Macke, A.: A new Saharan dust source activation frequency map derived from MSG-SEVIRI IR-channels, Geophys. Res. Lett., 34, L18803 (2007), https://doi.org/10.1029/2007GL030168
8. Seinfeld, J.H., Pandis, S.N.: Atmospheric Chemistry and Physics: From Air Pollution to Climate Change. John Wiley & Sons, Hoboken (2016)
9. Todaro, V., D'Oria, M., Tanda, M.G., Gómez-Hernández, J.J.: Ensemble smoother with multiple data assimilation for reverse flow routing. Comput. Geosci. **131**, 32–40 (2019)

10. Todaro, V., D'Oria, M., Tanda, M.G., Gómez-Hernández, J.J.: genES-MDA: a generic open-source software package to solve inverse problems via the ensemble smoother with multiple data assimilation. Comput. Geosci. **167**(12), 105210 (2022)
11. van Leeuwen, P.J., Evensen, G.: Data assimilation and inverse methods in terms of a probabilistic formulation. Mon. Weather Rev. **124**, 2898–2913 (1996)
12. Watson, J.G., Chow, J.C.: Reconciling urban fugitive dust emissions inventory and ambient source contribution estimates: summary of current knowledge and needed research. Technical report. Desert Research Institute, Document 6110.4F (2000)
13. Yu, Y., Kalashnikova, O.V., Garay, M.J., Lee, H., Notaro, M.: Identification and characterization of dust source regions across north Africa and the middle east using MISR satellite observations. Geophys. Res. Lett. **45**(13), 6690–6701 (2018)

Optimizing Pollution Control for an Economic Growth System

Vincenzo Capasso

Abstract In this chapter investigations are outlined on environmental issues concerning the control of the pollution produced by human activities. The proposed model consists of a spatially structured economic growth model which takes into account the level of pollution induced by production, a possible taxation based on the amount of produced pollution, and possible environmental interventions. It has been analyzed an optimal harvesting control problem with an objective function composed of four terms, namely the intertemporal utility of the decision maker, the space-time average of the level of pollution in the habitat, the disutility due to the imposition of taxation, and the cost of environmental interventions. A specific novelty in the model proposed here is the localization of the possible interventions to a subregion of the whole habitat. It clearly emerges that a reduction of the overall pollution leads to a significant improvement of the production.

Keywords Economic growth · Environmental quality · Control problems · Reaction-diffusion systems · Integral nonlocal term · Nonconcave production function

1 Introduction

Standard macroeconomics and environmental economics have been two completely independent research areas for a long time. The seminal paper by Grossman and Kruger [15] has stimulated considerable interest in the relationship between economic development and environmental pollution. The role of taxation for controlling pollution is the key issue of the report [17] issued by the International Monetary Fund. In view of these policy developments, some recent works tend to

V. Capasso (✉)
ADAMSS, Università degli Studi di Milano *La Statale*, Milano, Italy
e-mail: vincenzo.capasso@unimi.it

S. De Iaco et al. (eds.), *Exploration of Spatio-Temporal Environmental Conditions: Harmonized Databases and Analytical Techniques*, Springer Proceedings in Mathematics & Statistics 531, https://doi.org/10.1007/978-3-032-17526-7_7

develop a global theory combining the abovementioned two branches of literature [8, 11].

Our own results concern the impact of pollution diffusion on the economic and environmental joint dynamics and consider the optimal control problem of the relevant system subject to capital consumption, to taxation based on the amount of produced pollution, and possible environmental interventions. A key issue of our investigations is the localization of the possible interventions to a subregion of the whole habitat (in accordance with the motto *"Think globally, act locally"* which has been the *leit motiv* of our investigations for many years [4]).

In [1] and [2] the authors introduced a spatial economic growth model subject to pollution diffusion. In particular, in [1] the authors analyzed the large-time behavior of a spatially structured economic growth model coupling physical capital accumulation and pollution diffusion: this model extends other results in the literature along different directions. Furthermore, a negative feedback to the production function was added, in order to describe the (negative) influence of pollution on capital accumulation. In [2] the authors modified the model presented in [1] by introducing an optimal harvesting control problem by an objective function composed of three terms, namely the intertemporal utility of the decision-maker, the space-time average of the level of pollution in the habitat, and the disutility due to the imposition of taxation. In [3] two controls were introduced, namely the level of consumption and, in addition, the level of taxation on physical capital. This includes taxation on machinery, buildings, computers, and any equipment that help to turn the raw materials into finished products or services, and because of these activities they have a huge impact on pollution emissions.

The inclusion of a spatial dimension in economic analysis has recently increased relevance and interest. The first studies in economic geography go back to Beckerman [7] and Puu [18], who study regional problems based simply on flow equations. Starting from these works, a new economic geography arises adopting general equilibrium models to analyze the peculiarities of local and global markets and the mobility of production factors [14, 16]. More recently, this geographical approach has been introduced in economic growth models to study the implications between accumulation and diffusion of capital on economic dynamics [9, 10, 12].

The Solow model [19] with a continuous spatial dimension has been extensively studied. In our models the classical Solow model has been extended, by introducing a nonconcave production function. The Cobb-Douglas production function is by far the most used production function for describing situations with substitutional input factors although there are of course alternatives. Nonetheless, even if a Cobb-Douglas production function is not imposed, usually a production function f is assumed to be nonnegative, increasing, and concave and also fulfill the so-called Inada-Conditions [6] which read as follows:

$$\lim_{k \to 0} f'(k) = +\infty \tag{1}$$

Let us examine condition (1) from an economic point of view. We consider a region with nearly no physical capital, that is, there are no machines to produce goods with, no infrastructure, etc. This condition says that it is possible to gain infinitely high returns by investing only a small amount of money. This obviously cannot be realistic. Before getting returns it is necessary to create prerequisites by investing a certain amount of money. After establishing a basic structure for production, one might get only small returns until reaching a threshold where the returns will increase greatly to the point where the law of diminishing returns takes effect. In literature this fact is known as *poverty traps*. In other words, we should expect that there is a critical level of physical capital having the property that if the initial value of physical capital is smaller than such a level, then the dynamic of physical capital will descend to the zero level, thus vanishing any possibility of economic growth. Thus we have introduced the aforementioned threshold k^* and shift the whole function by this value

$$f(k) = \begin{cases} 0 & k < k^*; \\ f(k - k^*) & k \geq k^*. \end{cases} \tag{2}$$

A further generalization of the above model has been proposed in our literature [13], as follows:

$$f(k) = \frac{\alpha_1 k^\gamma}{1 + \alpha_2 k^\gamma}, \tag{3}$$

where $\alpha_1 \in (0, +\infty),\ \alpha_2 \in [0, +\infty),\ \gamma \in (0, +\infty)$.

In addition to the impact of taxation on the production as emerged in [3], in this work the role of additional costs due to possible environmental interventions analyzed in [5] has been reported.

This research project has been performed in collaboration with Sebastian Anita (Faculty of Mathematics, "Alexandru Ioan Cuza" University of Iasi, Iasi 700506, Romania) and Simone Scacchi (Department of Mathematics, Università degli Studi di Milano *La Statale*, 20133 Milano, Italy).

2 A Spatially Structured Model

By following our previous literature on the subject (see [3] and references therein), we shall consider the following spatially structured model. As far as the evolution of the stock capital is concerned, let $k(x, t)$ denote the capital stock held by the representative household located at $x \in \Omega$ (a non-empty bounded domain in $\mathbb{R}^n$, $n = 1, 2$, with a sufficiently smooth boundary), at time $t \geq 0$. The time evolution of the physical capital is modeled by

$$\frac{\partial k}{\partial t}(x, t) = d_1 \Delta k(x, t) + A(x, t) f\big(k(x, t)\big) - \delta_1 k(x, t) - c(x, t) k(x, t), \tag{4}$$

for $(x, t) \in \Omega \times (0, +\infty)$.

Here $A(x, t)$ denotes the technological progress available at $x \in \Omega$ and time $t \geq 0$. The parameter δ_1 takes into account the depreciation of the capital. $c(x, t)$ describes the level of consumption per unit of physical capital $k(x, t)$; the quantity $(1 - c(x, t))k(x, t)$ describes the saved capital after consumption.

Equation (4) is complemented by an initial capital distribution, $k(x, 0)$, and suitable boundary conditions. If we assume that there is no capital flow through the boundary $\partial\Omega$, we impose homogeneous Neumann boundary conditions, i.e.,

$$\frac{\partial k}{\partial n}(x, t) = 0, \quad x \in \partial\Omega, \ t > 0.$$

The production function f can be taken as anticipated in Eq. (3).

As far as pollution diffusion is concerned, if we denote by $p(x, t)$ the pollution stock faced by a representative household located at $x \in \Omega$, at time $t \geq 0$, we assume the following model:

$$\frac{\partial p}{\partial t}(x, t) = d_2 \Delta p(x, t) + \theta \int_{\Omega} f(k(x', t))\varphi(x', x)dx' - \delta_2 p(x, t). \tag{5}$$

The first term, including the Laplace operator, may model the random dispersal of the pollutant in the relevant habitat. Via the integral term in Eq. (5), we may see the impact of the economy on pollution production. At any point $x \in \Omega$ pollution is generated due to the production $f(k(x, t))$. In our model we have supposed that any point $x' \in \Omega$ may have an impact on enhancing pollution at any other point $x \in \Omega$, by any other possible mechanism than random dispersal already described by the Laplace operator. Such mechanisms are described by a kernel φ which satisfies the following hypotheses: $\varphi \in L^\infty(\Omega \times \Omega)$, and $\varphi(x', x) \geq 0$ a.e. $(x', x) \in \Omega \times \Omega$. As a limiting case one may reduce the kernel φ to a Dirac Delta function, to mean the impact of production on pollution at the same point x. The parameter θ takes into account the "efficiency" of the economic system with respect to pollution production. The parameter δ_2 accounts for the natural decay of pollution.

Equation (5) has to be complemented by suitable initial and boundary conditions.

Actually we have required that the model takes into account the negative feedback due to the impact of pollution on production, since it is expected that the relevant financial agents tend to reduce taxation and other indirect costs due to pollution, by a self-control; as a consequence in [3] we have modified Eq. (4) as follows:

$$\frac{\partial k}{\partial t}(x, t) = d_1 \Delta k(x, t) + \frac{A(x, t)}{g(p(x, t))} f(k(x, t)) - \delta_1 k(x, t) - c(x, t)k(x, t). \tag{6}$$

All parameters, $A, d_1, d_2, \delta_1, \delta_2$, which characterize a specific economic/ecological system are assumed strictly positive.

The function $g : [0, +\infty) \to (0, +\infty)$ models the negative impact of pollution on the production; usual assumptions on g are:

(i) g is continuously differentiable and monotonically increasing;
(ii) $g(0) = 1$ and $\lim_{p \to +\infty} g(p) = +\infty$.

As a typical choice for the function g, later we shall adopt the expression

$$g(p) = 1 + \chi p^2, \quad p \geq 0.$$

In the sequel it will be kept $\chi = 1$, as a typographical simplification.

3 An Optimal Control Problem

We plan to study our eco-economic system on a finite time horizon $[0, T]$. We shall denote by

$$Q_T = \Omega \times (0, T); \quad \Sigma_T = \partial\Omega \times (0, T).$$

In order to control pollution emissions by production processes, local authorities may impose taxes proportional to the level of production. A possible approach might be to impose taxes related to interventions on pollution abatement policies implemented only in the areas of interest, which we have indicate by a subregion $\omega \subset \Omega$ of the whole habitat. We shall denote $\tau(x, t)$ the environmental taxation rate. By taking all this into account, we need to modify Eqs. (5)–(6) as follows: [3, 11]:

$$\begin{cases} \dfrac{\partial k}{\partial t}(x,t) - d_1 \Delta k(x,t) + \dfrac{A(x,t)}{1 + p(x,t)^2} f((1 - \mathbb{I}_\omega(x)\tau(x,t))k(x,t)) \\ \qquad -\delta_1 k(x,t) - c(x,t)k(x,t), \\ \dfrac{\partial p}{\partial t}(x,t) = d_2 \Delta p(x,t) + \theta \displaystyle\int_\Omega f((1 - \mathbb{I}_\omega(x')\tau(x',t))k(x',t))\varphi(x',x)dx' \\ \qquad -\delta_2 p(x,t) - \mathbb{I}_\omega(x)\xi(x,t)p(x,t), \end{cases} \tag{7}$$

for $(x, t) \in Q_T$, where ω is a nontrivial open subset of Ω. System (7) shall be complemented by initial and boundary conditions to be precise later.

The term $(1 - \tau(x, t))k(x, t)$ accounts for the amount of the physical capital left to be used in the gross domestic production. An increase in the level of taxation allows the implementation of practices leading to an abatement of the level of pollution emissions, so that we may interpret it as a "green" taxation policy. By $\xi(x, t)$ we have denoted the rate of pollution abatement by suitable environmental interventions.

In this case the optimal control problem would be

$$\underset{\substack{(c,\tau,\xi)\in\widetilde{U},\\ \omega\subset\Omega}}{Maximize} \left\{ \int_0^T \int_\Omega c(x,t)k(x,t)dx\,dt - \beta_0 \int_0^T \int_\Omega p(x,t)dx\,dt \right.$$

$$- \beta_1 \int_0^T \int_\omega \tau(x,t)k(x,t)dx\,dt$$

$$\left. -\beta_2 \int_0^T \int_\omega \xi(x,t)p(x,t)dx\,dt \right\},$$

subject to System (7) and the initial and boundary conditions given below.

It is clear that the solutions of the above problems would depend upon the geographical distribution of the potentially polluting production activities; in this respect we may expect a crucial role played by the functional parameter $A(x,t)$. This fact would confirm the relevance of spatially structured models for geographical economics.

For simplicity we may consider homogeneous Neumann boundary conditions

$$\frac{\partial k}{\partial n}(x,t) = \frac{\partial p}{\partial n}(x,t) = 0, \qquad (x,t) \in \Sigma_T, \tag{8}$$

and initial conditions

$$k(x,0) = k_0(x), \; p(x,0) = p_0(x), \qquad x \in \Omega, \tag{9}$$

where $T > 0$ is fixed and $k_0, p_0 \in L^\infty(\Omega)$, $k_0(x) \geq 0$, $p_0(x) \geq 0$ a.e. $x \in \Omega$.

We impose the constraints

$$1 - c(x,t) - \tau(x,t) \geq s, \;\; c(x,t), \tau(x,t) \geq 0, \quad \text{a.e. } (x,t) \in Q_T, \tag{10}$$

where $s > 0$ is the saving factor. In addition, we may assume that $0 \leq \xi(x,t) \leq L$, a.e. $(x,t) \in \omega \times (0,T)$, for a given constant L.

If we denote $U = \{(c,\tau) \in L^\infty(Q_T) \times L^\infty(Q_T) : \; (c,\tau) \text{ satisfies Eq. (10)}\}$, the set of controls is $\widetilde{U} := U \times \{\xi \in L^\infty(\omega \times (0,T)) : \; 0 \leq \xi(x,t) \leq L, \text{ a.e. } (x,t) \in \omega \times (0,T)\}$, and $\beta_0, \beta_1, \beta_2$ are positive constants; (k,p) is the solution to (7)–(9) corresponding to Q_T and Σ_T. The objective function is composed of three terms that we hereby describe:

$T1$: the term $\int_0^T \int_\Omega c(x,t)k(x,t)dx\,dt$, to be maximized, describes the intertemporal utility of the decision-maker,

$T2$: the second term $\beta_0 \int_0^T \int_\Omega p(x,t)dx\,dt$, to be minimized, represents an average of the total level of pollution across space and time,

$T3:$ the third term $\beta_1 \int_0^T \int_\Omega \tau(x,t)k(x,t)dx\,dt$, to be minimized, takes into account the disutility due to imposition of taxation,

$T4:$ the fourth term $\beta_2 \int_0^T \int_\Omega \xi(x,t)p(x,t)dx\,dt$, to be minimized, takes into account the costs due to environmental interventions.

Since the search of the optimal control is not an easy analytical task, we have proposed a computational algorithm for the search by the gradient method.

Numerical simulations have been carried out for various scenarios, showing a clear agreement with our intuition; in particular it clearly emerges that a reduction of the overall pollution leads to a significant increase of the production [5].

Bibliography

1. Aniţa, S., Capasso, V., Kunze, H., La Torre, D.: Optimal control and long-run dynamics for a spatial economic growth model with physical capital accumulation and pollution diffusion. Appl. Math. Lett. **26**, 908–912 (2013)
2. Aniţa, S., Capasso, V., Kunze, H., La Torre, D.: Dynamics and optimal control in a spatially structured economic growth model with pollution diffusion and environmental taxation. Appl. Math. Lett. **42**, 36–40 (2015)
3. Aniţa, S., Capasso, V., Kunze, H., La Torre, D.: Optimizing environmental taxation on physical capital for a spatially structured economic growth model including pollution diffusion. Vietnam J. Math. **45**, 199–206 (2017)
4. Aniţa, S., Capasso, V., Scacchi, S.: Mathematical Modeling and Control in Life and Environmental Sciences – Regional Control Problems. Springer, Berlin (2024)
5. Aniţa, S., Capasso, V., Scacchi, S.: Pollution control for a spatially structured economic growth system (2025). https://doi.org/10.48550/arXiv.2510.27393
6. Barro, R.J., Sala-I-Martin, X.: Economic Growth. MIT Press, Cambridge (2004)
7. Beckerman, W.: Economic growth and the environment: Whose growth? Whose environment? World Develop. **20**(4), 481–496 (1992)
8. Bertinelli, L., Strobl, E., Zou, B.: Sustainable economic development and the environment: theory and evidence. Energy Econ. **34**(4), 1105–1114 (2012)
9. Boucekkine, R., Camacho, C., Zou, B.: Bridging the gap between growth theory and the new economic geography: the spatial Ramsey model. Macroecon. Dyn. **13**(4), 20–45 (2009)
10. Brito, P.: The dynamics of growth and distribution in a spatially heterogeneous world. Technical report, UECE-ISEG, Technical University of Lisbon (2004)
11. Brock, W.A., Taylor, M.S.: The green Solow model. J. Econ. Growth **15**(2), 127–153 (2010)
12. Camacho, C., Zou, B.: The spatial Solow model. Econ. Bull. **18**(2), 1–11 (2004)
13. Capasso, V., Engbers, R., La Torre, D.: On the spatial Solow model with technological diffusion and nonconcave production function. Nonlinear Anal. Real World Appl. **11**(2), 3858–3876 (2010)
14. Fujita, M., Krugman, P., Venables, A.: The Spatial Economy. Cities, Regions and International Trade. MIT Press, Cambridge (1999)
15. Grossman, G.M., Krueger, A.B.: Economic growth and the environment. Q. J. Econ. **110**(2), 353–377 (1995)
16. Krugman, P.: Increasing returns and economic geography. J. Polit. Econ. **99**(3), 483–499 (1991)

17. Norregard, J., Reppelin-Hill, V.: Taxes and tradable permits as instruments for controlling pollution: theory and practice. Technical report. IMF Working paper WP/00/13 (2000), http://www.imf.org/external/pubs/ft/wp/2000/wp0013.pdf
18. Puu, T.: Outline of a trade cycle model in continuous space and time. Geogr. Anal. **14**(1), 1–9 (1982)
19. Solow, R.M.: A contribution to the theory of economic growth. Q. J. Econ. **70**(1), 65–94 (1956)

Exploring Functional Structure and Variation of Italian Carbon Emissions

Pierdomenico Duttilo and Tonio Di Battista

Abstract This work explores carbon emissions from electricity generation in Italy between 2021 and 2023 using functional data analysis. Daily emission profiles are constructed from hourly data using Fourier basis functions. The analysis covers both the national level and the electricity market zones, highlighting key differences in emission levels and variability. In addition, the functional principal component analysis is applied to find the main patterns of variation, the first two components explaining more than 96% of the total variability in Italian emissions. This study provides valuable information to support energy policies and emission reduction strategies.

Keywords Carbon emissions · Electricity generation · Functional data analysis · Functional principal component analysis

1 Introduction

Addressing climate change requires a substantial reduction in greenhouse gas (GHG) emissions. Electricity generation is one of the main sources of global GHG emissions [8]. The amount of carbon emissions (CO_2) released depends on the type of energy used: renewable sources such as solar and wind energy and fossil fuels such as coal, oil and gas [10]. Reducing CO_2 emissions from electricity generation is essential to achieve international climate targets, including those outlined in the Paris Agreement of 2015, which aims to limit the increase in global temperatures to well below 2 °C [15].

P. Duttilo (✉)
Department of Economics, Engineering, Society and Business, University of Tuscia, Viterbo, Italy
e-mail: pierdomenico.duttilo@unitus.it

T. Di Battista
Department of Socio-Economic, Management and Statistical Studies, University "G. d'Annunzio" of Chieti-Pescara, Pescara, Italy

S. De Iaco et al. (eds.), *Exploration of Spatio-Temporal Environmental Conditions: Harmonized Databases and Analytical Techniques*, Springer Proceedings in Mathematics & Statistics 531, https://doi.org/10.1007/978-3-032-17526-7_8

This work aims to explore Italian carbon emissions in the power sector during the period 2021–2023 using functional data analysis (FDA). Given the continuous nature of carbon emission processes, the FDA provides a more natural framework for analysing emission data [3, 12]. CO_2 emissions are examined both at the aggregate level and in individual market zones of the Italian electricity market: North, Centre-North, Centre-South, South, Calabria, Sicily and Sardinia. In addition, a functional principal component analysis (FPCA) is performed to investigate the main components of the variation in Italian carbon emissions.

The remainder of the chapter is structured as follows: Sect. 2 describes the CO_2 emissions data and outlines the functional descriptive statistics and FPCA. Section 3 presents the results and provides concluding remarks.

2 Data and Methodology

Carbon emissions from electricity generation at time i (E_i) are calculated using the approach based on emission factors [2], defined as

$$\begin{aligned} E_i &= \sum_{f=1}^{F} E_{i,f}, \\ E_{i,f} &= G_{i,f} \times EF_f \times O_f \times M, \end{aligned} \tag{1}$$

where $E_{i,f}$ represents the carbon emissions at time i associated with fuel type f; $G_{i,f}$ denotes the hourly electricity generation by fuel type f at time i, obtained from the European network of transmission system operators for electricity [6]; EF_f is the country-specific emission factor for fuel type f, sourced from the Italian institute for environmental protection and research; O_f corresponds to the country-specific oxidation rate for fuel type f [9]; and M is the molecular weight ratio of CO_2 to carbon, fixed at $44/12 = 3.6667$.

The dataset covers hourly observations from 1 January 2021 to 31 December 2023 and includes both national-level and zonal-level data for the following electricity market zones: North, Centre-North, Centre-South, South, Calabria, Sicily and Sardinia. Figure 1 illustrates the hourly CO_2 emissions from electricity generation (in tCO_2) in Italy between 1 January 2021 and 31 December 2023.

2.1 From Raw Data to Functional Data

Let $\mathcal{H}$ be a separable Hilbert space, and consider the continuous-time carbon emissions process $E_i(\tau)$ defined in $\mathcal{H}$ in the time domain $\tau \in \mathrm{T}$. In this work $E_i(\tau) \in L^2[1, 24]$ with i indicating a given daily profile for $i = 1, 2, \ldots, N$. $E_i(\tau)$

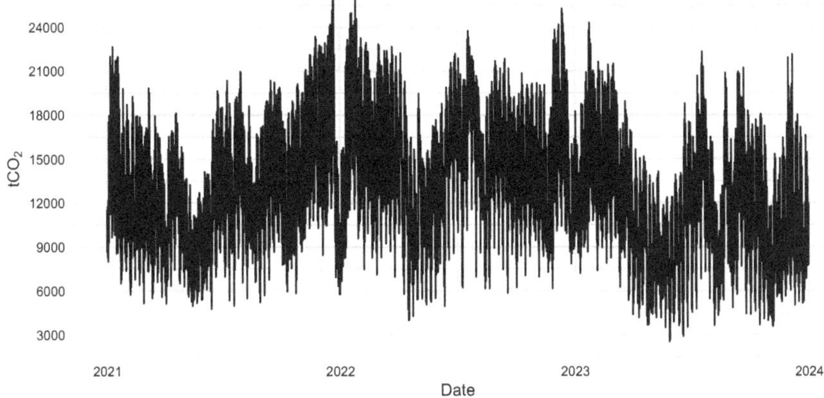

Fig. 1 Hourly CO_2 emissions from electricity generation in Italy for the period 2021–2023

can be expanded as a linear combination of a set of basis functions ϕ_k:

$$E_i(\tau) = \sum_{k=1}^{K} c_{ik}\phi_k(\tau), \ \text{for } i = 1, 2, \ldots, N. \tag{2}$$

Specifically, hourly carbon emissions data are transformed into daily curves using Fourier expansion, which is well suited for modelling periodic data. The orthonormal version of the Fourier basis in the space $L^2[1, 24]$ is given by

$$\begin{aligned} \phi_0(\tau) &= \frac{1}{T^{1/2}} \\ \phi_{2k-1}(\tau) &= \left(\frac{2}{T}\right)^{1/2} \sin\left(\frac{2\pi k\tau}{T}\right) \\ \phi_{2k}(\tau) &= \left(\frac{2}{T}\right)^{1/2} \cos\left(\frac{2\pi k\tau}{T}\right), \ \text{for } k = 1, \ldots, K. \end{aligned} \tag{3}$$

Hourly carbon emissions are transformed into daily curves by smoothing the 24 hourly observations of each day using a 12-Fourier basis. Figure 2 compares the raw and smoothed carbon emissions for the first week of 2023 (Sunday, 1 January–Saturday, 7 January) in Italy. The red, pink and violet curves highlight the lower emission levels typically observed on weekends and bank holidays.

Figure 3 shows the smoothed monthly emission curves for Italy by year, along with the overall functional mean (dashed black line). Monthly emission curves have been obtained by aggregating hourly data and then smoothed using 9-Fourier basis. They peak in winter and then gradually decrease in spring. A moderate increase is observed in summer, mainly due to air conditioning use, followed by a decrease in

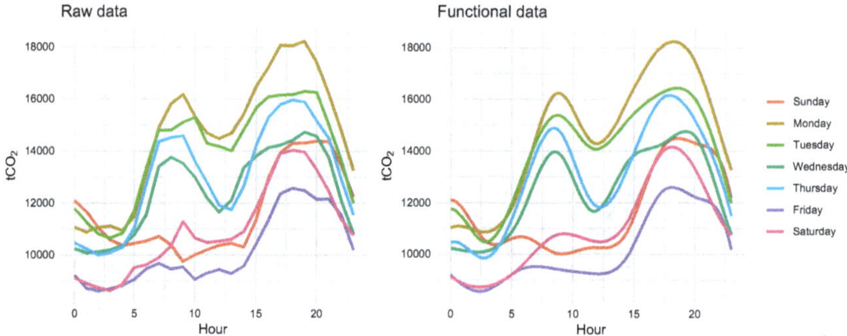

Fig. 2 Raw and functional data for carbon emissions, first week 2023, Italy

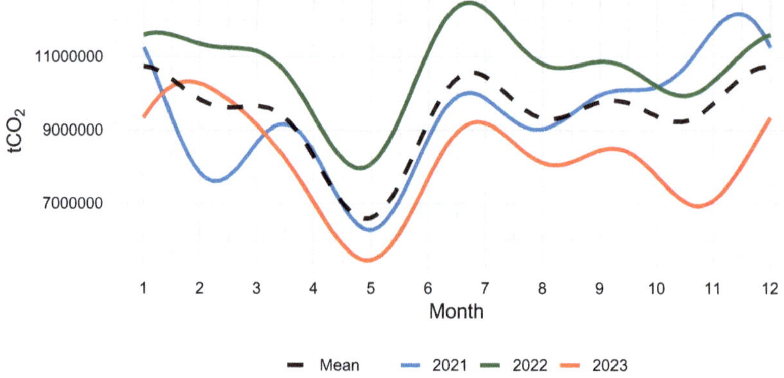

Fig. 3 Monthly functional carbon emission curves in Italy by year; the overall functional mean in dashed black line

August during bank holidays. Finally, they increase again in autumn. The emission level in 2022 increased compared to 2021, likely as a consequence of the outbreak of the war in Ukraine [7]. However, in 2023, emissions decreased relative to the previous 2 years.

2.2 Functional Descriptive Statistics

Let $\{E_i(\tau), \tau \in [1, 24], i = 1, \dots, N\}$ be a sample of N daily carbon emission curves; the sample mean function can be defined as follows:

$$\bar{E}(\tau) = \frac{1}{N} \sum_{i=1}^{N} E_i(\tau), \tag{4}$$

where $\bar{E}(\tau)$ is an unbiased estimator of the mean function $\mu(\tau)$, $\mathbb{E}[\bar{E}(\tau)] = \mu(\tau)$. The sample variance function is defined as follows:

$$s^2(\tau) = \frac{1}{N-1} \sum_{i=1}^{N} [E_i(\tau) - \bar{E}(\tau)]^2. \tag{5}$$

The bivariate covariance function specifies the correlation between the curve values $E_i(\tau)$ and $E_i(h)$ at times h and τ, respectively. It is estimated by

$$c(h, \tau) = \frac{1}{N-1} \sum_{i=1}^{N} [E_i(h) - \bar{E}(h)][E_i(\tau) - \bar{E}(\tau)]. \tag{6}$$

The correlation function instead is estimated as follows:

$$r(h, \tau) = \frac{cov(h, \tau)}{\sqrt{s^2(\tau)s^2(h)}}. \tag{7}$$

2.3 *Functional Principal Component Analysis*

The FPCA is a functional extension of the conventional principal component analysis. The purpose of FPCA is to represent the main modes of variation in the sample through a set of orthogonal functions, known as functional principal components [12, 13]. The j-th functional principal component is a linear combination of the original variable with maximum variance

$$z_{ij} = \int_{1}^{24} E_i(\tau)\, w_j(\tau)\, d\tau, \tag{8}$$

where $w_j(\tau)$ is the j-th principal component weight function. This function is chosen to maximise the variance of the corresponding score z_{ij}, subject to orthogonality constraints with the previously estimated components. The functions $w_j(\tau)$ are estimated from the data and provide insight into the dominant patterns of variation in the emission curve sample.

3 Explorative Analysis and FPCA Results

The functional exploratory analysis of daily carbon emissions from electricity generation in Italy between 2021 and 2023 shows how emissions vary during the day and differ between zones of the electricity market.

Figure 4 displays the functional mean of carbon emissions by zones. All regions exhibit a typical daily pattern characterised by two distinct peaks, one in the morning and another in the evening, corresponding to periods of higher electricity demand. However, the amplitude and the shape of these peaks, as well as the midday trough between them, vary between zones. The evening peak is generally more pronounced than the morning peak, especially in the South, Sicily and Sardinia. In general, the North, Centre-South and South show the highest average emission levels, while the Centre-North, Calabria, Sicily and Sardinia report lower emissions.

Figure 5 illustrates the functional standard deviation, capturing the hourly variability in carbon emissions. In Italy and within all zones, variability generally increases around morning and evening peaks, indicating greater fluctuations in electricity generation during periods of high demand. Calabria, Sicily and Sardinia exhibit moderate levels of variability, whereas the Centre-North shows consistently lower variability during the day.

Figure 6 displays the bivariate correlation function, which captures the correlation of carbon emissions between different hours. For Italy and the North, the surfaces appear broad and relatively flat at the top, indicating consistently high correlation values throughout the day. In contrast, the surfaces for Sicily and Sardinia exhibit steeper gradients, suggesting that correlations decline more rapidly as the time gap between hours increases.

The FPCA was applied to the daily carbon emission curves of Italy from 2021 to 2023. Two functional principal components were selected, together explaining 96.9% of the total variation. As shown in Fig. 7, the first principal component accounts for 92.5% of the variation and closely captures the variance function shown in Fig. 5. The second component, which accounts for 4.4% of the variation, reflects both the timing of peak emissions and the contrast between emission levels during peak and off-peak periods within the daytime interval from 8:00 to 19:00.

Although this work aims to explore the CO_2 emissions from electricity generation, it can be used to support energy policies taking into account the different zones of the Italian electricity market. Future extensions could include the use of functional principal component analysis for short-term forecasting of CO_2 emissions [5, 11], by using the Karhunen–Loève decomposition [1, 14] and accounting for the dependence among curves [4].

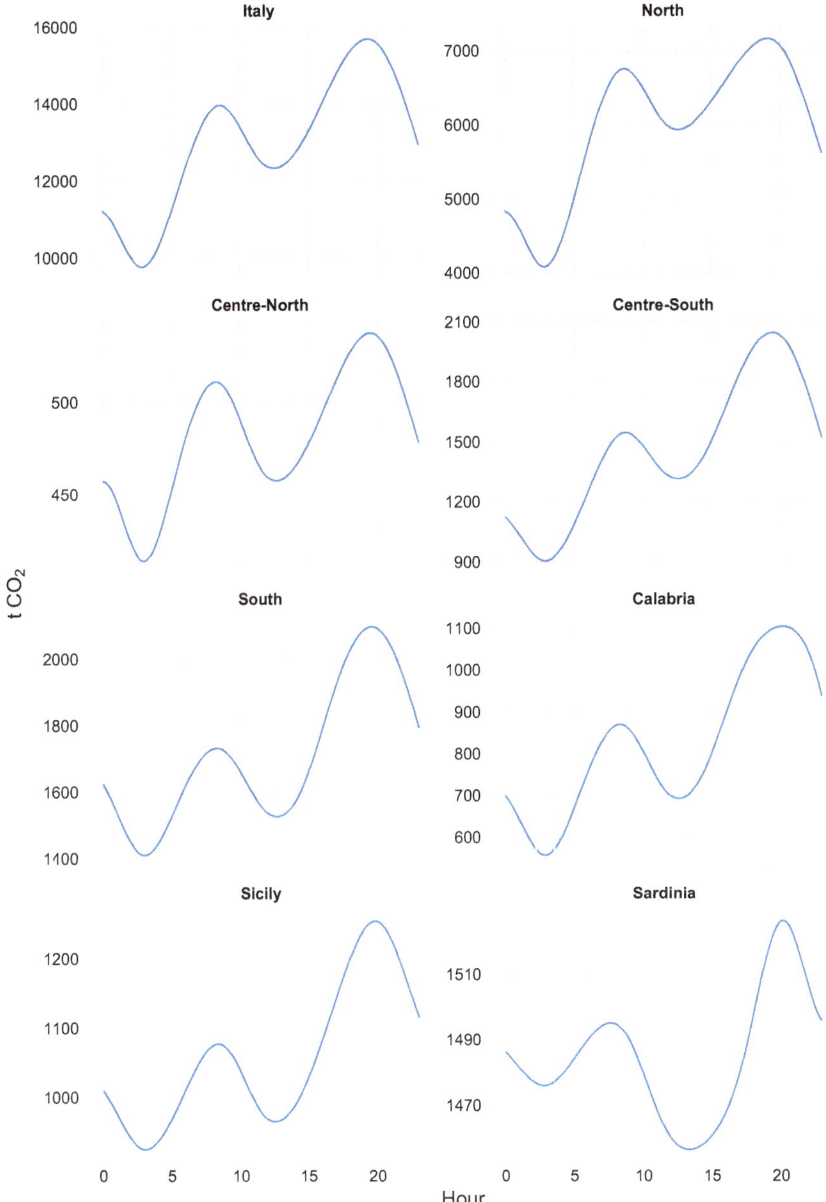

Fig. 4 Functional mean of daily carbon emissions in Italy and its electricity market zones from 2021 to 2023

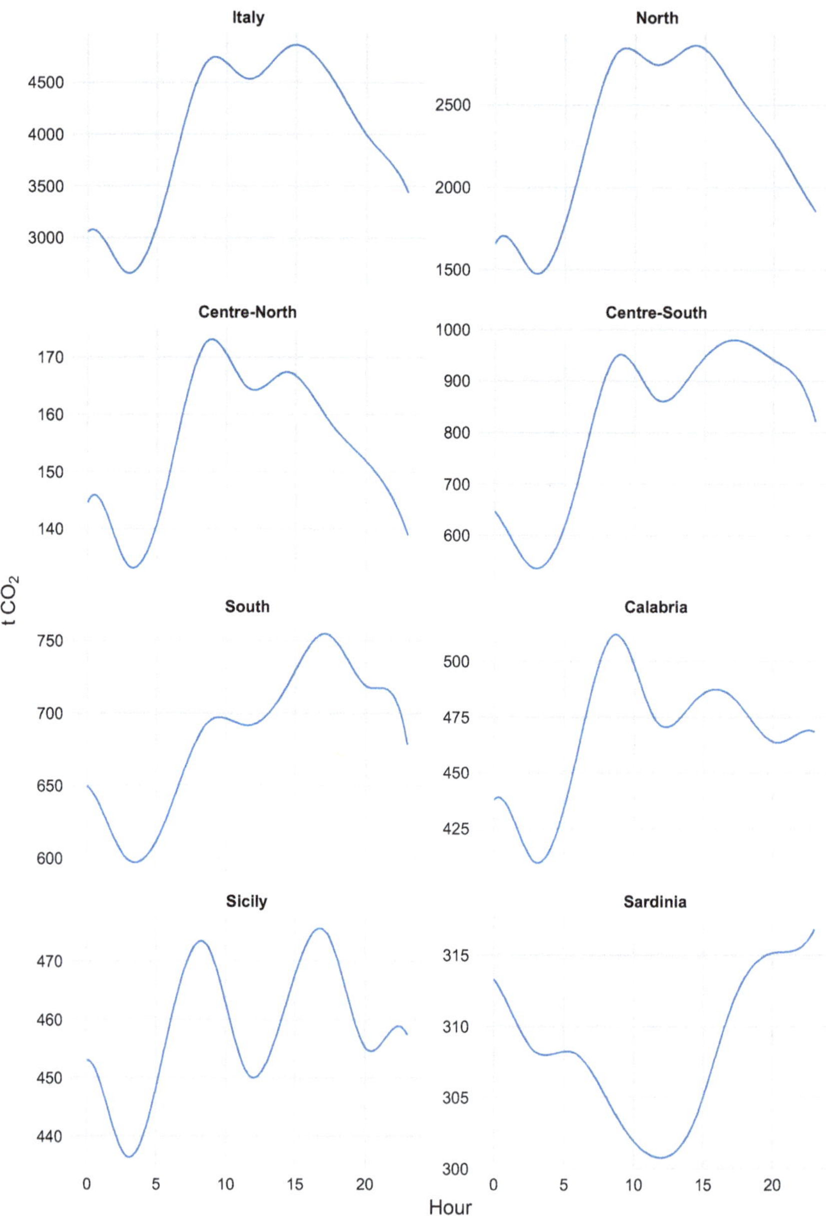

Fig. 5 Functional standard deviation of daily carbon emissions in Italy and its electricity market zones from 2021 to 2023

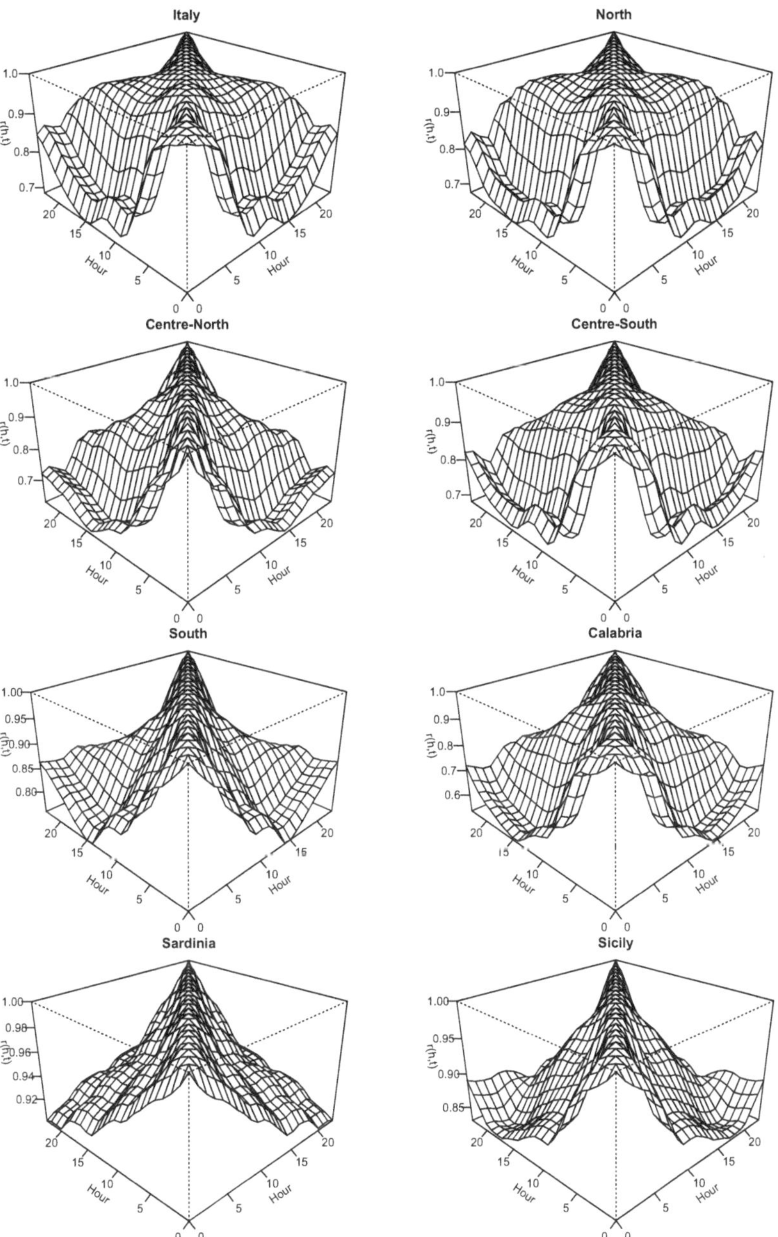

Fig. 6 Functional bivariate correlation of daily carbon emissions in Italy and its electricity market zones from 2021 to 2023

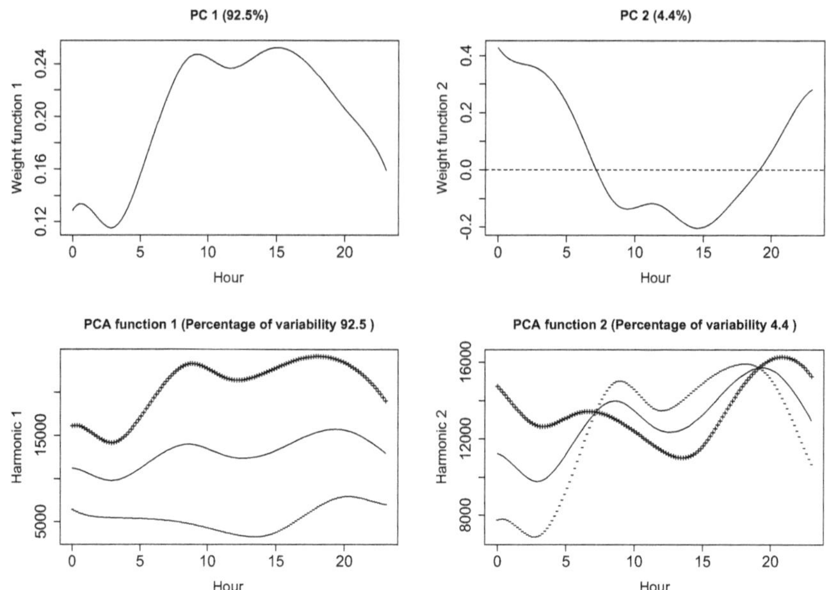

Fig. 7 The top panel displays the first two principal component functions. Below, their effects are illustrated as perturbations of the mean function (solid line): '+' denotes the result of adding a small amount of the principal component to the mean, while '−' represents the effect of subtracting it

Bibliography

1. Aguilera, A., Ocana, F., Valderrama, M.: Forecasting time series by functional PCA. Discussion of several weighted approaches. Comput. Stat. **14**(3), 443–467 (1999)
2. Bertolini, M., Duttilo, P., Lisi, F.: Accounting carbon emissions from electricity generation: a review and comparison of emission factor-based methods. Appl. Energy **392**, 125992 (2025)
3. Chen, Y., Koch, T., Lim, K.G., Xu, X., Zakiyeva, N.: A review study of functional autoregressive models with application to energy forecasting. WIREs Comput. Stat. **13**(3), e1525 (2021)
4. Damon, J., Guillas, S.: The inclusion of exogenous variables in functional autoregressive ozone forecasting. Environmetrics **13**(7), 759–774 (2002)
5. Duttilo, P., Lisi, F.: Forecasting carbon emissions from electricity generation: classical vs functional methods. In: IES 2025 Innovation & Society Statistics and Data Science for Evaluation and Quality-BOOK OF SHORT PAPERS, pp. 91–98. CLEUP sc (2025)
6. ENTSO-E: Actual generation per production type. https://www.entsoe.eu/ (2025). Accessed 03 Jan 2025
7. Friedrich-Ebert-Stiftung (FES): Just transition in Europe: Challenges and opportunities for coal regions. https://library.fes.de/pdf-files/bueros/budapest/20542.pdf (2025). Accessed 23 June 2025
8. IEA: Global CO_2 emissions by sector. https://www.iea.org/data-and-statistics/charts/global-co2-emissions-by-sector-2017 (2025). Accessed 23 June 2025
9. ISPRA: Emission factors for the production and consumption of electricity in Italy. https://emissioni.sina.isprambiente.it (2025). Accessed 03 Jan 2025

10. Kocak, E., Ulug, E.E., Oralhan, B.: The impact of electricity from renewable and non-renewable sources on energy poverty and greenhouse gas emissions (GHGs): empirical evidence and policy implications. Energy **272**, 127125 (2023)
11. Leerbeck, K., Bacher, P., Junker, R.G., Goranović, G., Corradi, O., Ebrahimy, R., Tveit, A., Madsen, H.: Short-term forecasting of CO2 emission intensity in power grids by machine learning. Appl. Energy **277**, 115527 (2020)
12. Ramsay, J., Silverman, B.W.: Functional Data Analysis. Springer Series in Statistics. Springer, New York (2006)
13. Ramsay, J., Hooker, G., Graves, S.: Functional Data Analysis with R and MATLAB. Springer, New York (2009)
14. Shang, H.L., Hyndman, R.J.: Nonparametric time series forecasting with dynamic updating. Math. Comput. Simul. **81**(7), 1310–1324 (2011)
15. UNFCCC: The Paris Agreement. https://unfccc.int/process-and-meetings/the-paris-agreement (2025). Accessed 23 June 2025

Part III
Environmental Analysis

Ozone Predictions Through a Generalized Additive Model

Abdollah Jalilian, Claudia Cappello, Monica Palma, and Sandra De Iaco

Abstract Accurate prediction of surface-level ozone (O_3) concentrations is essential for ensuring efficient air quality control and safeguarding public health. This study proposes a flexible spatiotemporal modeling framework to estimate daily mean O_3 levels across Italy by integrating satellite-derived ozone data with high-resolution environmental predictors and in situ observations. The model relies on a linear regression with dynamic intercepts and slopes to relate in situ O_3 measurements to satellite estimates, accounting for additive and multiplicative biases. The time- and space-varying parameters are modeled within a generalized additive model (GAM) framework to capture complex relationships between ozone and environmental drivers. Model evaluation and cross-validation demonstrate superior explanatory and predictive performance compared to other alternatives. The framework identifies key bias patterns linked to elevation, nitrogen dioxide concentrations, and seasonal variation. Overall, the proposed model provides an accurate, scalable, and interpretable framework for obtaining harmonized databases, facilitating environmental monitoring and informing policy development.

A. Jalilian
Lancaster Ecology and Epidemiology Group, Lancaster University, Lancaster, UK

C. Cappello (✉)
Department of Economic Sciences, University of Salento, Lecce, Italy
e-mail: claudia.cappello@unisalento.it

M. Palma
Department of Economic Sciences, University of Salento, Lecce, Italy

National Centre for HPC, Big Data and Quantum Computing, Bologna, Italy

S. De Iaco
Department of Economic Sciences, University of Salento, Lecce, Italy

National Centre for HPC, Big Data and Quantum Computing, Bologna, Italy

National Biodiversity Future Center, Palermo, Italy

S. De Iaco et al. (eds.), *Exploration of Spatio-Temporal Environmental Conditions: Harmonized Databases and Analytical Techniques*, Springer Proceedings in Mathematics & Statistics 531, https://doi.org/10.1007/978-3-032-17526-7_9

Keywords Generalized additive model · Surface-level ozone · Harmonized databases

1 Introduction

Accurate estimation and prediction of surface-level exposure to ozone (O_3) are essential for managing air quality and protecting public health and food security, given that ambient O_3 poses serious risks to human health and negatively affects crop quality and yield [24, 34]. Interpolation techniques and predictive modeling approaches in air quality management generally include physics-based photochemical grid models (e.g., CAMx, as discussed in [10]), data-driven machine learning models (as in [8, 19, 25, 28]), statistical models (as in [12, 13, 30]), even referred to spatiotemporal data (as in [4–7, 26]), or based on dimensionality reduction techniques (as in [2, 21, 22, 27]).

In this context, statistical modeling provides an essential framework for deriving interpretable connections between O_3 levels and environmental variables, along with explicit measures of uncertainty [12]. Land use regression (LUR) models, which typically employ a multiple linear regression framework, are commonly used to relate pollutant concentrations to spatially resolved geographic and environmental predictors [29]. To capture complex and potentially nonlinear spatiotemporal patterns, Generalized Additive Models (GAMs), firstly introduced by Hastie [14, 15], have been widely used in air pollution research, allowing for smooth variation across time [20] and space-time [16, 31].

This study proposes an additive regression model with spatiotemporal varying coefficients to relate daily mean in situ O_3 concentrations to satellite-derived O_3 estimates across Italy. Similarly to LUR models, this approach incorporates a comprehensive set of environmental variables within an additive framework to capture variations in both intercept and slope coefficients in space and time. It is worth noting that the proposed additive structure improves the flexibility of the model, allowing it to capture the potentially nonlinear effects of the covariates within the linear relationship. After describing the dataset used in the case study (Sect. 2), the modeling framework has been introduced in Sect. 3. The model's goodness of fit and its spatial and temporal predictive capabilities have been evaluated through an extensive cross-validation procedure (Sect. 4). The article concludes with the main findings and implications (Sect. 5).

2 Dataset

In this study, ground-based O_3 concentrations measured at Italian air quality monitoring stations have been used as the primary response variable. Explanatory variables have comprised satellite-derived O_3 estimates and environmental predictors representing atmospheric composition, meteorological conditions, topography, land use, and anthropogenic activity. Data spanning a 10-year period, from January 1, 2013 to December 30, 2023, have been included in the analysis.

2.1 *Ozone Monitoring Data (Response Variable)*

Hourly O_3 concentrations have been obtained from the European Air Quality Portal maintained by the European Environment Agency (EEA) [11]. In particular, data for 402 Italian air quality monitoring stations have been considered and then aggregated to calculate daily mean O_3 concentrations for the 10-year period. Mean daily O_3 values have exhibited spatiotemporal variability and a right-skewed distribution, characterized by frequent moderate and infrequent high concentrations. To mitigate skewness and approximate normality, a square-root transformation has been applied, consistently with prior studies [9, 18] and supported by Box–Cox analysis [1]. Monthly boxplots of transformed O_3 concentrations have revealed a distinct seasonal cycle, peaking in summer due to meteorological factors like sunlight and temperature [33].

2.2 *Covariates*

A comprehensive set of spatiotemporal covariates from satellite as well as static covariates has been employed in the analysis:

1. *CAMS Air Quality Data:* daily mean concentrations of O_3, NO_2, $PM_{2.5}$, and PM_{10} from the Copernicus Atmosphere Monitoring Service (CAMS) European reanalysis data at a $0.1°$ resolution (approximately 10 km) [3]. The square-root-transformed CAMS O_3 concentration has shown a strong linear relationship with the in situ data (Pearson correlation coefficient of 0.9139). Note that the empirical variogram of the differences between CAMS and station values has indicated no spatial structure in the discrepancies.
2. *ERA5-Land Climate Variables:* daily means of meteorological variables, including air temperature at 2m, surface net solar radiation, relative humidity, and boundary layer height, from the ERA5-Land dataset [23].
3. *Spatial Environmental/Anthropogenic Variables:* elevation, population density (proxy for anthropogenic emissions), road density (indicator of traffic-related emissions), and land cover/use (influencing biogenic emissions and chemistry).

3 Generalized Additive Model Framework

Let $Y(\mathbf{u}, t)$ be the stochastic process associated with the square root of the mean daily O_3 concentration at location $\mathbf{u} \in \mathbb{R}^2$ and day $t \in \mathbb{Z}$ and $z(c_0(\mathbf{u}), t)$ be the corresponding CAMS O_3 concentration at the day t and the satellite grid cell $c_0(\mathbf{u})$ that contains the location $\mathbf{u}$. The spatiotemporal process $Y(\mathbf{u}, t)$ has been modeled as

$$Y(\mathbf{u}, t) = \mu(\mathbf{u}, t) + \alpha(\mathbf{u}, t) z(c_0(\mathbf{u}), t) + \epsilon(\mathbf{u}, t), \tag{1}$$

where:

- $\mu(\mathbf{u}, t)$ is a spatially and temporally varying intercept that accounts for systematic differences between satellite-derived and in situ O_3 observations,
- $\alpha(\mathbf{u}, t)$ represents a dynamic slope, enabling the relationship between satellite and in situ O_3 measurements to vary in space and time,
- $\epsilon(\mathbf{u}, t)$ corresponds to an unstructured random error component, assumed to be independent and identically distributed (i.i.d.) following a Student's t-distribution with $\nu > 2$ degrees of freedom.

To enhance interpretability, $\mu(\mathbf{u}, t)$ and $\alpha(\mathbf{u}, t)$ have been related to p environmental covariates (x_j) using an additive structure, forming the full model (M_3):

$$\mu(\mathbf{u}, t) = \beta_0 + \sum_{j=1}^{p} f_j(x_j(c_j(\mathbf{u}), t)) + s(d(t)) + \tau_{w(t)} \tag{2}$$

$$\alpha(\mathbf{u}, t) = \alpha_0 + \sum_{j=1}^{p} \tilde{f}_j(x_j(c_j(\mathbf{u}), t)) + \tilde{s}(d(t)) + \tilde{\tau}_{w(t)}. \tag{3}$$

In this Generalized Additive Model (GAM) structure:

- f_j and $\tilde{f}_j$ are smooth functions, implemented using thin plate regression splines, as in [32], capturing potentially nonlinear effects of covariates,
- $s(d(t))$ and $\tilde{s}(d(t))$ are penalized cyclic cubic regression splines, as in [32], based on the day of the year $d(t)$, capturing seasonal variation,
- $\tau_{w(t)}$ and $\tilde{\tau}_{w(t)}$ are Gaussian i.i.d. random effects for the day of the week $w(t)$, accounting for potential weekly patterns like the "weekend ozone effect."

Parameter estimation is achieved efficiently using the restricted maximum likelihood approach via the `mgcv` package in R [32].

4 Model Evaluation and Results

Model M_3, with both intercept and slope modeled as additive nonlinear functions with seasonal and weekly components, has been compared with simpler nested models. In particular, the following alternative models have been considered:

- M_0, with constant intercept $\mu(\mathbf{u}, t) = \beta_0$ and slope $\alpha(\mathbf{u}, t) = \alpha_0$

- M_1, with linear varying intercept $\mu(\mathbf{u}, t) = \beta_0 + \sum_{j=1}^{p} \beta_j x_j(\mathbf{u}, t)$ and constant slope $\alpha(\mathbf{u}, t) = \alpha_0$
- M_2, with seasonal and weekly components $\mu(\mathbf{u}, t) = \beta_0 + \sum_{j=1}^{p} f_j\left(x_j(\mathbf{u}, t)\right) + s(d(t)) + \tau_{w(t)}$ and constant slope $\alpha(\mathbf{u}, t) = \alpha_0$

The inclusion of additive nonlinear structures and dynamic slope coefficients in M_3 has led to a consistent and statistically significant improvement in all goodness of fit metrics (R^2_{adj}, %Dev, AIC, and BIC) compared to M_0, M_1, and M_2, as confirmed by partial F-statistics ($p < 0.001$ for all comparisons).

The full model M_3 has achieved an adjusted coefficient of determination (R^2_{adj}) of 0.867 and explained 72.5% of the null deviance. Residual diagnostics have supported the assumption of the Student's t-distribution ($\hat{\nu} = 3.6$) and showed no clear structure against fitted values, supporting homoscedasticity.

4.1 Model Interpretability of Dynamic Coefficients

The dynamic coefficient framework, proposed in Eq. (1), has successfully quantified how specific environmental variables influence the additive bias ($\mu(\mathbf{u}, t)$) and the multiplicative bias ($\alpha(\mathbf{u}, t)$).

- *Elevation* has shown a substantial positive effect on $\mu(\mathbf{u}, t)$, associated with a nonlinear upward shift (up to 6 μg/m^3 on the original scale) in satellite estimates relative to in situ observations at higher elevations. In contrast, the influence of elevation on multiplicative bias $\alpha(\mathbf{u}, t)$ was negligible.
- *Nitrogen Dioxide* (NO_2) has shown a positive effect on $\mu(\mathbf{u}, t)$ (upward bias) but a negative effect on $\alpha(\mathbf{u}, t)$ (slight scaling contraction). This implies that high NO_2 conditions are linked to an upward bias in satellite estimates alongside a reduction in the strength of their association with ground measurements.
- *Seasonal Variation (Day of Year)* has affected both $\mu(\mathbf{u}, t)$ and $\alpha(\mathbf{u}, t)$, indicating that seasonal factors influence both the baseline offset and the scaling relationship between the two measurement sources.

4.2 Predictive Performance

Predictive performance has been assessed using three cross-validation (CV) schemes: General k-fold, Spatial CV (testing unobserved locations using k-medoid clustering), and Temporal CV (forecasting using only past data).

In particular, in the general CV each observation has been randomly assigned to one of k mutually exclusive groups, yielding a partition of the dataset into subsets $\mathcal{P}_1, \ldots, \mathcal{P}_k$. For each fold $j = 1, \ldots, k$, the model has been fitted using all observations except those contained in $\mathcal{P}_j$, after which predictions are generated for the excluded subset. This general k-fold cross-validation procedure provides an integrated evaluation of the model's predictive capability and its generalization performance across data partitions.

In the spatial CV approach, the 402 air quality monitoring stations have been grouped into k spatial clusters, $\mathcal{C}_1, \ldots, \mathcal{C}_k$, using the k-medoids algorithm [17]. These clusters define spatially coherent partitions of the dataset into subsets $\mathcal{P}_1, \ldots, \mathcal{P}_k$, where each subset $\mathcal{P}_j$ contains all observations from the stations belonging to cluster $\mathcal{C}_j$ across the full 4016-day study period. For each fold $j = 1, \ldots, k$, the observations in $\mathcal{P}_j$ are omitted, the model has been trained on data from the remaining $k - 1$ subsets, and O_3 concentrations are subsequently predicted for the held-out subset. This spatial cross-validation framework is designed to evaluate the model's capability to predict O_3 concentrations at unobserved locations and to assess its spatial generalization performance.

Finally, in the temporal CV the full study period comprising 4016 days has been divided into $k + 1$ consecutive time intervals of equal length, denoted as $\mathcal{I}_0, \mathcal{I}_1, \ldots, \mathcal{I}_k$. The dataset has been accordingly partitioned into temporal subsets $\mathcal{P}_0, \ldots, \mathcal{P}_k$, where each subset $\mathcal{P}_j$ includes all observations from every monitoring station corresponding to days within interval $\mathcal{I}j$. For each fold $j = 1, \ldots, k$, the model is trained using data from the immediately preceding subset $\mathcal{P}_{j-1}$, and predictions of O_3 concentrations have been generated for the subsequent subset $\mathcal{P}_j$. This temporal cross-validation approach emulates realistic forecasting conditions by restricting model training to past data, thereby preventing temporal information leakage and enabling a rigorous assessment of the model's predictive accuracy over time.

Model predictive performance for each cross-validation strategy has been quantified using the following accuracy metrics [31]: the Root Mean Square Prediction Error (RMSPE), the Mean Absolute Prediction Error (MAPE), and the cross-validated coefficient of determination (CVR^2), defined as follows:

$$\text{RMSPE} = \frac{1}{k} \sum_{j=1}^{k} \sqrt{\frac{1}{\#(\mathcal{P}_j)} \sum_{(\mathbf{u},t)\in\mathcal{P}_j} \left(Y^2(\mathbf{u}, t) - \widehat{Y^2}(\mathbf{u}, t)\right)^2},$$

$$\text{MAPE} = \frac{1}{k} \sum_{j=1}^{k} \left(\frac{1}{\#(\mathcal{P}_j)} \sum_{(\mathbf{u},t)\in\mathcal{P}_j} \left| Y^2(\mathbf{u}, t) - \widehat{Y^2}(\mathbf{u}, t) \right| \right),$$

$$\text{CVR}^2 = \frac{1}{k} \sum_{j=1}^{k} \left(1 - \frac{\sum_{(\mathbf{u},t)\in\mathcal{P}_j} \left(Y(\mathbf{u}, t) - \widehat{Y}(\mathbf{u}, t)\right)^2}{\sum_{(\mathbf{u},t)\in\mathcal{P}_j} \left(Y(\mathbf{u}, t) - \frac{1}{\#(\mathcal{P}_j)} \sum_{(\mathbf{u},t)\in\mathcal{P}_j} Y(\mathbf{u}, t)\right)^2} \right),$$

where $Y^2(\mathbf{u}, t)$ represents the observed O_3 concentration at site $\mathbf{u}$ and day t, and $\widehat{Y^2}(\mathbf{u}, t)$ denotes the predicted one. The notation $\#(\mathcal{P}_j)$ corresponds to the number of data points included in subset $\mathcal{P}_j$. Note that since RMSPE squares the prediction

Table 1 Cross-validation performance metrics (in bold the best predictive performance)

	Spatial CV			Temporal CV			General CV		
Model	RMSPE	MAPE	CVR^2	RMSPE	MAPE	CVR^2	RMSPE	MAPE	CVR^2
M_0	12.024	6.572	0.738	11.862	6.225	0.814	11.880	6.133	0.818
M_1	11.438	6.268	0.763	11.124	5.922	0.837	11.046	5.792	0.842
M_2	11.446	6.391	0.762	10.871	5.813	0.844	10.683	5.615	0.853
M_3	**11.382**	**6.306**	**0.763**	**10.744**	**5.714**	**0.857**	**10.505**	**5.472**	**0.867**

errors, it emphasizes larger discrepancies, whereas MAPE expresses the mean absolute deviation and is generally easier to interpret.

To assess how the varying intercept term (2) and the dynamic slope term (3) contribute to predictive performance, the full model M_3 has been evaluated against three simplified alternatives (M_0–M_2) in predicting O_3 concentrations, $Y^2(\mathbf{u}, t)$. A summary of the results has been provided in Table 1, which reports RMSPE, MAPE, and CVR^2 values under the spatial, temporal, and general cross-validation frameworks. Under both temporal and general CV, the full model M_3 exhibit superior predictive performance, achieving the lowest RMSPE and MAPE, and the highest CVR^2. Predicting O_3 concentrations at unobserved spatial locations (Spatial CV) has been confirmed to be more challenging, yielding higher errors compared to temporal CV. Moreover, M_3 achieved the lowest RMSPE (11.382) in Spatial CV, demonstrating superior overall accuracy by reducing the magnitude of larger prediction errors when generalizing to novel locations.

5 Conclusions

This research successfully established a flexible and interpretable modeling framework for surface-level O_3 prediction by employing a Generalized Additive Model with spatiotemporally varying intercept and dynamic slope coefficients. The novelty lies in combining the established GAM methodology with dynamic coefficient modeling, enabling a clear and unified mechanism to understand and adjust for both additive and multiplicative biases in satellite-based O_3 measurements relative to in situ observations. The results from goodness-of-fit metrics and cross-validation strongly demonstrated that incorporating these dynamic components substantially improved model fit and predictive accuracy compared to models assuming constant coefficients. Moreover, by revealing how environmental factors drive the relationship between in situ and satellite observations, the model contributes to obtaining harmonized data, supporting environmental monitoring and evidence-based policy development.

Bibliography

1. Box, G.E., Cox, D.R.: An analysis of transformations. J. R. Stat. Soc. Ser. B: Stat. Methodol. **26**(2), 211–243 (1964)
2. Cabral Pinto, F., Manchuk, J.G., Deutsch, C.V.: Decomposition of multivariate spatial data into latent factors. Comput. Geosci. **153**, 104773 (2021)
3. Copernicus Atmosphere Monitoring Service: CAMS European air quality reanalyses (2021). Accessed on 10 Apr 2025
4. De Iaco, S., Myers, D.E., Posa, D.: Space-time variograms and a functional form for total air pollution measurements. Comput. Stat. Data Anal. **41**(2), 311–328 (2002)
5. De Iaco, S., Myers, D.E., Posa, D.: The linear coregionalization model and the product-sum space-time variogram. Math. Geol. **35**(1), 25–38 (2003)
6. De Iaco, S., Myers, D.E., Palma, M., Posa, D.: Using simultaneous diagonalization to identify a space-time linear coregionalization model. Math. Geosci. **45**(1), 69–86 (2013)
7. De Iaco, S., Palma, M., Posa, D.: Choosing suitable linear coregionalization models for spatio-temporal data. Stochastic Environ. Res. Risk Assess. **33**(7), 1419–1434 (2019)
8. De Iaco, S., Hristopulos, D.T., Lin, G.: Special issue: geostatistics and machine learning. Math. Geosci. **54**(3), 459–465 (2022)
9. Dou, Y., D. Le, N., V. Zidek, J.: Modeling hourly ozone concentration fields. Ann. Appl. Stat. **4**(3), 1183–1213 (2010)
10. Emery, C.A., Baker, K.R., Wilson, G.M., Yarwood, G.: Comprehensive air quality model with extensions, v7. 20: formulation and evaluation for ozone and particulate matter over the US. Geosci. Model Dev. Discuss. **2024**, 1–48 (2024)
11. European Environment Agency: European air quality portal. https://www.eea.europa.eu/themes/air/air-quality-index (2025). Accessed 10 Apr 2025
12. Gelfand, A.E., Fuentes, M., Hoeting, J.A., Smith, R.L.: Handbook of Environmental and Ecological Statistics. CRC Press, Boca Raton (2019)
13. Genton, M., Kleiber, W.: Cross-covariance functions for multivariate geostatistics. Stat. Sci. **30**(2), 147–163 (2015)
14. Hastie, T., Tibshirani, R.: Generalized additive models. Stat. Sci. **1**(3), 297–310 (1986)
15. Hastie, T.J., Tibshirani, R.J.: Generalized Additive Models. Chapman and Hall, London (1990)
16. Jalilian, A., Cappello, C., Palma, M., De Iaco, S.: Generalized Additive Model With Dynamic Coefficients for Spatiotemporal Ozone Predictions. Environmetrics **37**(3), (2026)
17. Kaufman, L., Rousseeuw, P.J.: Finding Groups in Data: An Introduction to Cluster Analysis. John Wiley & Sons, Hoboken (1990)
18. Lu, X., Gelfand, A.E., Holland, D.M.: Local real-time forecasting of ozone exposure using temperature data. Environmetrics **29**(7), e2509 (2018)
19. Meng, X., Wang, W., Shi, S., Zhu, S., Wang, P., Chen, R., Xiao, Q., Xue, T., Geng, G., Zhang, Q., et al.: Evaluating the spatiotemporal ozone characteristics with high-resolution predictions in mainland China, 2013–2019. Environ. Pollut. **299**, 118865 (2022)
20. Morlini, I., Albertson, S., Orlandini, S.: Characterization of annual urban air temperature changes with special reference to the city of Modena: a comparison between regression models and a proposal for a new index to evaluate relationships between environmental variables. Stochastic Environ. Res. Risk Assess. **38**(3), 1163–1178 (2024)
21. Muehlmann, C., De Iaco, S., Nordhausen, K.: Blind recovery of sources for multivariate space-time random fields. Stochastic Environ. Res. Risk Assess. **37**, 1593–1613 (2023)
22. Muehlmann, C., Cappello, C., De Iaco, S., Nordhausen, K.: Anisotropic local covariance matrices for spatial blind source separation. AStA Adv. Stat. Anal. (2025)
23. Muñoz Sabater, J.: ERA5-Land hourly data from 1950 to present (2019). Accessed on 10 Apr 2025
24. Nuvolone, D., Petri, D., Voller, F.: The effects of ozone on human health. Environ. Sci. Pollut. Res. **25**(9), 8074–8088 (2018)

25. Pan, Q., Harrou, F., Sun, Y.: A comparison of machine learning methods for ozone pollution prediction. J. Big Data **10**(1), 63 (2023)
26. Rouhani, S., Wackernagel, H.: Multivariate geostatistical approach to space-time data analysis. Water Resour. Res. **26**(4), 585–591 (1990)
27. Sipilä, M., Cappello, C., De Iaco, S., Nordhausen, K., Taskinen, S.: Modelling multivariate spatio-temporal data with identifiable variational autoencoders. Neural Netw. **181**, 106774 (2025)
28. Sipilä, M., Cappello, C., De Iaco, S., Nordhausen, K., Palma, M., Taskinen, S.: Enhancing identifiable variational autoencoder in the presence of missing values and auxiliary covariates for spatio-temporal ozone modeling (2025). Submitted
29. van Nunen, E., Vermeulen, R., Tsai, M.Y., Probst-Hensch, N., Ineichen, A., Davey, M., Imboden, M., Ducret-Stich, R., Naccarati, A., Raffaele, D., et al.: Land use regression models for ultrafine particles in six European areas. Environ. Sci. Technol. **51**(6), 3336–3345 (2017)
30. Wackernagel, H.: Multivariate Geostatistics: an Introduction with Applications. Springer, Berlin (2003)
31. Wang, J., Cohan, D.S., Xu, H.: Spatiotemporal ozone pollution LUR models: suitable statistical algorithms and time scales for a megacity scale. Atmos. Environ. **237**, 117671 (2020)
32. Wood, S.N.: Generalized Additive Models: an Introduction with R. Chapman and Hall/CRC, Boca Raton (2017)
33. Yan, Y., Pozzer, A., Ojha, N., Lin, J., Lelieveld, J.: Analysis of European ozone trends in the period 1995–2014. Atmos. Chem. Phys. **18**(8), 5589–5605 (2018)
34. Zhang, J., Wei, Y., Fang, Z.: Ozone pollution: a major health hazard worldwide. Front. Immunol. **10**, 2518 (2019)

Characterization and Forecasting of Drought: Geostatistical Approaches

Miguel Gomes, Ana Castro, and Amilcar Soares

Abstract Droughts represent one of the most significant natural hazards, with far-reaching impacts, especially in vulnerable regions such as Southern Portugal. This study investigates long-term changes in drought characteristics across this region over the past four decades, focusing on severity, spatial patterns, and variability. Two meteorological indices—Consecutive Dry Days (CDD) and RL10 (days with <10 mm precipitation)—were analyzed for the periods 1980–2000 and 2001–2020 using data from 17 meteorological stations. Findings reveal a marked intensification of droughts and rising spatial uncertainty. A Joint Drought Index (JDI), combining both indicators through multidimensional scaling, highlights a northward expansion of drought. To support early detection and mitigation of drought events, a Random Forest-based short-term alert system capable of accurately forecasting drought severity classes at the monthly scale was developed. These findings offer a flexible tool that can be applied to other regions, contributing to global efforts in managing drought-related risks.

Keywords Drought trends · Consecutive Dry Days (CDD) · Joint Drought Index (JDI) · Spatial uncertainty · Machine learning prediction

1 Objectives

Droughts are natural phenomena that significantly contribute to desertification, leading to soil and water loss, ecosystem degradation, and environmental migration.

M. Gomes · A. Soares (✉)
CERENA/Department of Mineral and Energy Resources Engineering, Instituto Superior Técnico, Lisboa, Portugal
e-mail: asoares@tecnico.ulisboa.p

A. Castro
CERENA, ISEP, Polytechnic of Porto, Porto, Portugal

S. De Iaco et al. (eds.), *Exploration of Spatio-Temporal Environmental Conditions: Harmonized Databases and Analytical Techniques*, Springer Proceedings in Mathematics & Statistics 531, https://doi.org/10.1007/978-3-032-17526-7_10

In recent decades, climate change has intensified the frequency and severity of drought events.

Droughts have had a profound impact across the entire Mediterranean Basin. However, this study specifically examines the evolving dynamics of drought in southern Portugal over the last four decades, focusing on changes in severity, variability, and spatial patterns. Meteorological indices, including Consecutive Dry Days (CDD) and the number of days with precipitation below 10 mm (RL10), are analyzed across two consecutive 20-year periods: 1980–2000 and 2001–2020. In its final stage, this study proposes an alert system based on a machine learning (ML) short-term predictor for drought severity.

2 Study Region and Drought Indices

This study relies on precipitation data from the Portuguese Institute for Sea and Atmosphere (IPMA). The dataset includes daily precipitation records from 17 meteorological stations (MSs) located in the southern regions of mainland Portugal (Fig. 1), with most stations providing data spanning from 1980 to 2020.

2.1 Drought Indices—RL10 and CDD

The RL10 index measures the number of days in a year with less than 10 mm of precipitation [2, 6, 11].

The CDD index measures the number of consecutive days with less than 1 mm of precipitation. This drought index is endorsed by the Expert Team on Climate

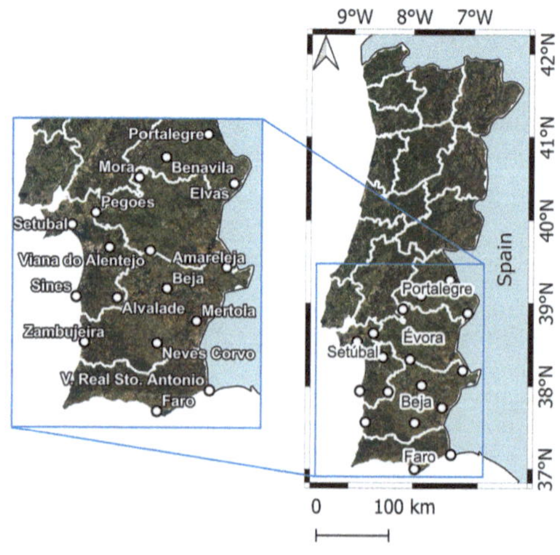

Fig. 1 Study area and monitoring stations' locations

Change Detection and Indices (ETCCDI) [21] and is widely used in drought studies [15, 18, 19].

3 Spatial Characterization of the Main Temporal Trends

3.1 Drought Temporal Trends

To assess drought temporal trends and local spatial patterns in southern Portugal, the 40-year period (1980–2020) was divided into two equal-length intervals: T1 (1980–2000) and T2 (2001–2020) [7].

Histograms of CDD events were generated for each monitoring station across periods T1 and T2. Box plots of Fig. 2 show a clear trend from T1 to T2 across all key statistics (Q1, median, Q3, and maximum) at every monitoring station.

3.2 Spatial Characterization of Local Trends of the Median, Q3, and IQR

To identify critical local areas based on the trends of the previously defined drought indices, spatial models of median, third quartile Q3, and interquartile range (IQR) were generated using geostatistical simulation. Instead of using direct sequential

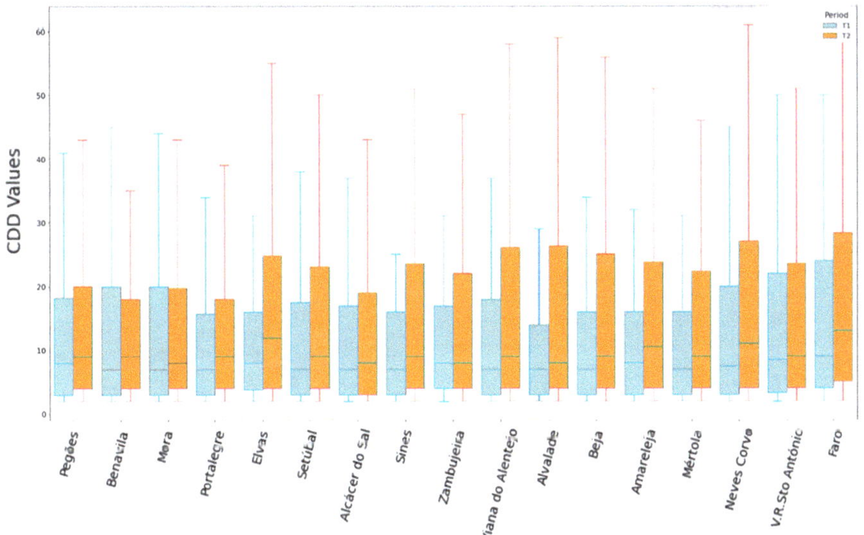

Fig. 2 Box plot of CDD values for each MS during T1 (1980–2000) and T2 (2001–2020)

simulation with local probability distributions [9] to characterize local distributions at unsampled locations based on known CDD histograms at monitoring stations (as presented in Sect. 3.1), we opted for a simpler approach. This method directly characterizes the median, third quartile, and interquartile range—parameters known at the monitoring stations—which can be spatially represented using stochastic simulation, following the interpolation of histograms formalism: the Quantile Interpolation model [8, 10, 13], which is based on the following approach: A local quantile $Q_z(x_0)$ of a probability density function (PDF) located at unsampled location x_0 can be obtained by simple kriging [5], based on known quantiles $Q_z(x_\alpha)$ located at the surrounding monitoring stations x_α (1).

$$Q_z\,(x_0) = \sum_{\alpha=1}^{N} \lambda_\alpha(x_0) Q_z(x_\alpha) \tag{1}$$

This local mean and the local variance identified with the simple kriging variance are the basic parameters to draw a simulated value $Q_z^l(x_0)$, by using direct sequential simulation [17].

Direct sequential simulation was conducted for three variables—median, Q3, and IQR. Figure 3 presents the average models from 32 realizations of the median and Q3 for both periods, T1 and T2, illustrating the spatial representation of the temporal trend of these statistics over the last 20 years (period T2).

3.3 *Temporal Trend of Spatial Uncertainty of Drought Phenomenon*

The interquartile range (IQR) of the ensemble of simulated realizations serves as a measure of spatial uncertainty in CDD. Figure 4 presents the mean IQR model for periods T1 and T2, illustrating an increase in spatial uncertainty in T2. This trend

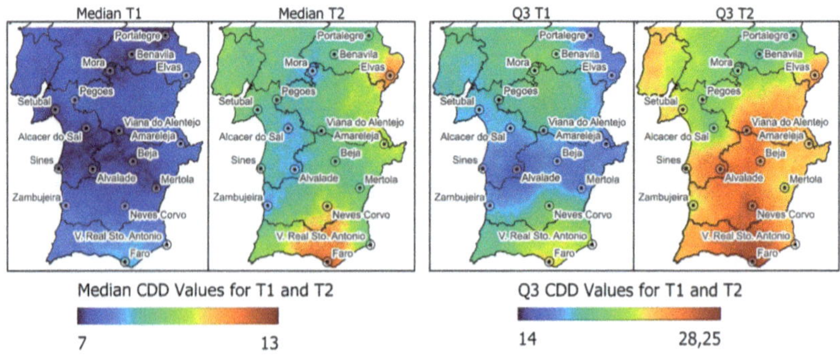

Fig. 3 Spatial distribution of CDD median and Q3 values for T1 and T2

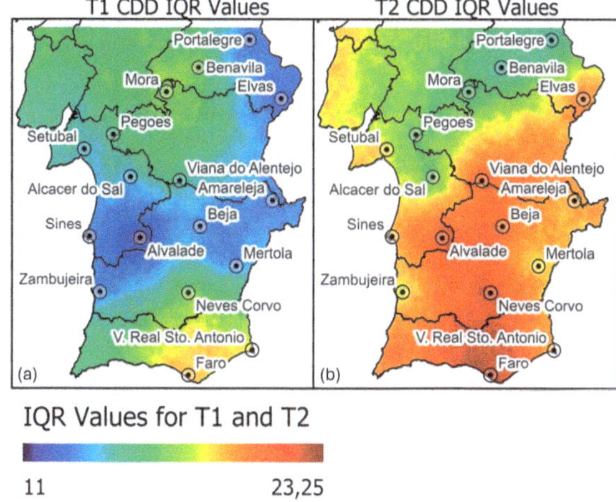

Fig. 4 Spatial distribution of IQR for CDD in T1 (**a**) and T2 (**b**)

suggests that the drought phenomenon is not only intensifying (as shown in Fig. 3) but also becoming more locally uncertain, making it progressively less predictable over time.

3.4 Drought Risk Classification

Drought risk classes were established using statistical percentiles derived from stochastic simulations. Table 1 summarizes the threshold values for each risk class, based on the third quartile (Q3) of the full ensemble of monitoring stations for periods T1 and T2. For instance, in each simulated realization, a spatial location x_0 is classified as the most severe, Class 4, if its simulated Q3 value exceeds the global 90th percentile (P90).

Following *N* stochastic simulations, the most frequently observed drought risk class in each pixel was selected to generate the final risk maps for T1 and T2. Figure 5 presents these drought risk maps, revealing a significant increase in the Extreme and High drought risk classes in T2. Specifically, areas exceeding the P75 threshold (Classes 3 and 4) expanded by 50% from T1 to T2, while regions surpassing the P90 threshold (Class 4) grew by 25%.

Table 1 Description of drought risk classes

Class	Description	Percentile range
Class 4	Extreme drought risk	Above P90
Class 3	High drought risk	Between P75 and P90
Class 2	Moderate drought risk	Between P50 and P75
Class 1	Low drought risk	Below P50

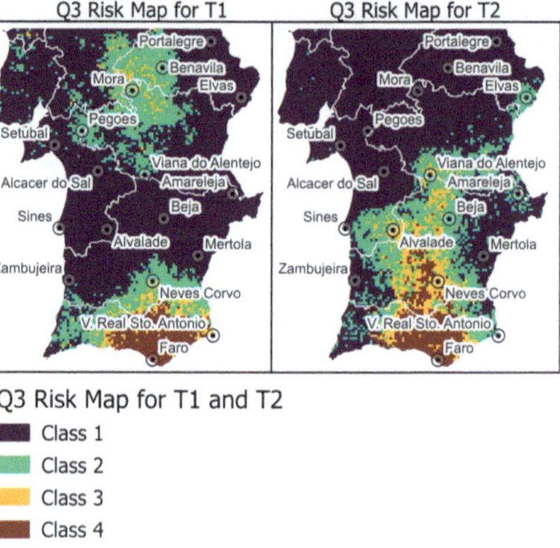

Fig. 5 Evolution of drought risk (Q3) from T1 to T2

3.5 *Joint Drought Index*

The drought indicators CDD and RL10 complement each other. The CDD index quantifies the continuity of dry spells, while the RL10 index captures low-precipitation events. To integrate these two indicators, a dimensionality reduction technique was used. Several authors have used geostatistical approaches with dimensionality reduction in a space-time context [4, 14]. In this study, a Joint Drought Index (JDI) was developed using a multivariate dimensionality reduction approach, specifically Multidimensional Scaling (MDS), as described by Borg and Groenen [1], Cox and Cox [3], Scheidt and Caers [16], Torgerson [20].

For this exercise, the following severity drought classes, based on CDD, were adopted (Table 2)

The proportion of each class $p_c(x, t)$ was determined for each monitoring station located at x in periods T1 and T2:

$$p_c(x, t) = \frac{N_c(x, t)}{N(x, t)} \tag{2}$$

Table 2 Description of drought risk classes

Class	Description	CCD
Class A	Extreme drought events with dry spells exceeding 70 consecutive days	CDD > 70
Class B	Severe droughts with durations between 20 and 70 consecutive days	$20 \leq$ CDD ≤ 70
Class C	Moderate droughts, characterized by dry periods lasting less than 20 consecutive days	CDD < 20

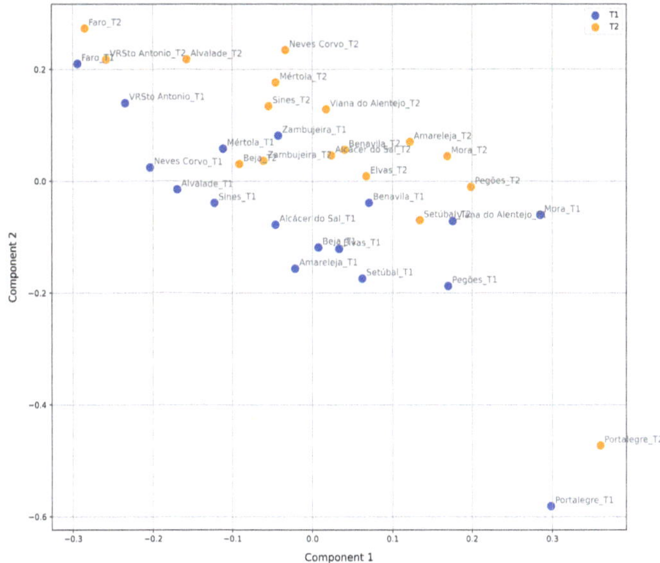

Fig. 6 MDS plot of T1 and T2 MS

where $N_c(x,t)$ is the number of days in each class and $N(x,t)$ is the total number of CDDs in the same period and monitoring station.

In the MDS analysis, the dissimilarity between pairs of monitoring stations is determined by the proportions of drought classes (as defined in Sect. 3.2), along with the RL10 average index for each period at both stations. Figure 6 presents the results of MDS, depicting the set of monitoring stations in T1 and T2 within a two-dimensional space. Both components (axes) indicate a consistent shift from T1 to T2; however, component 2 appears to more effectively capture the severity trend of the drought phenomenon as represented by the two indicators.

We generated spatial dispersion maps of the Joint Drought Index using the same methods outlined in previous sections, but with "MDS Component 2" as the JDI variable. Figure 7 presents the average results from 32 simulations in T1 and T2, revealing even clearer local trends over time. Notably, this new variable highlights the northward expansion of the drought phenomenon from T1 to T2.

4 An Alert System for Short-Term Prediction of Drought Severity

One of the key findings of this study is that drought events—especially the most severe—are becoming increasingly spatially uncertain, making them more unpredictable over time. This challenge led to the final phase of the research:

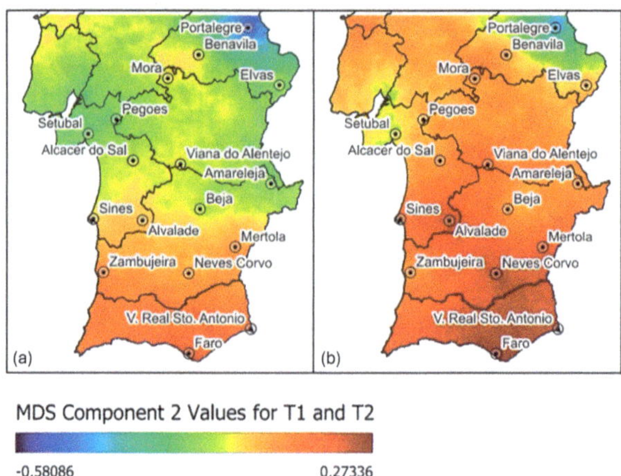

Fig. 7 Average of the DSS applied to the Y-MDS values for T1 (**a**) and T2 (**b**)

designing a short-term forecast alert system for drought severity. Here, "short-term" refers to predictions within the same hydrological season, specifically on a monthly scale.

This phase aims to evaluate the effectiveness of the Random Forest (RF) model in predicting drought severity, using Consecutive Dry Days (CDD) as the primary drought indicator, specifically focusing on the three drought severity classes defined in Sect. 3.5.

The target output is the drought severity class at a given monitoring station, x_0, for a short-term forecast at time t_0. The prediction considers the severity class at the same location over the previous 2 months. Additionally, it incorporates the average severity class for the target month t_0 from previous years at the same monitoring station as an input.

The short-term input variable is the drought severity class, measured by the number of days in categories A, B, and C (defined in Sect. 3.5) over the 2 months preceding the target month t_0. The long-term input variable accounts for the number of days in each drought severity class (A, B, and C) for the target month t_0 across the previous 3 years. Finally, the output represents the CDD classification A, B, or C that the RF model seeks to predict for the target month t_0 at each monitoring station.

For training, 80% of the total dataset was randomly selected, while the remaining 20% was used to test and evaluate the model's predictive performance. To develop the RF models, several parameters were optimized using the GridSearchCV tool from the scikit-learn library [12] to identify the best-performing configuration for each model.

To visualize the spatial distribution of predicted drought severity, we selected the predicted and actual values from a specific period in July for illustrative purposes. We conducted the same simulation exercise outlined in Sect. 3.2, but this time incorporating the real and predicted values for each class. We conducted

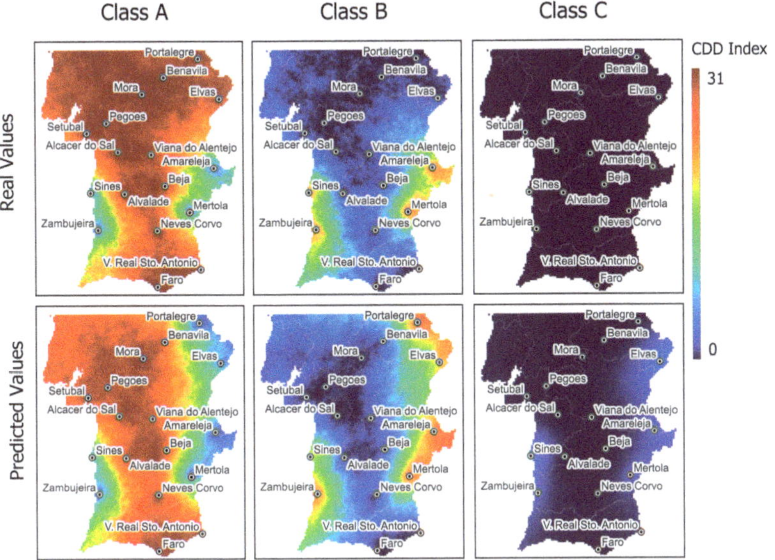

Fig. 8 Average maps of simulated realizations of real and predicted values for drought severity classes A, B, and C

32 realizations of DSS, with the mean values represented in Fig. 8. The strong predictive performance for local areas is evident when compared to the actual values.

Based on the ensemble of simulated realizations, we identify the most extreme and critical local areas by generating maps of the probability of exceedance (PE) beyond the 75th percentile (P75) for the predicted values of Classes A and B (Fig. 9). These maps highlight the areas most likely to surpass the defined critical thresholds.

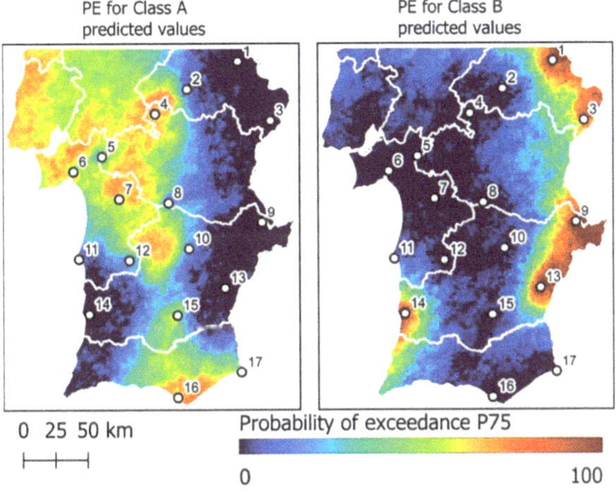

Fig. 9 Probability of exceedance of P75 for Classes A and B predicted values

5 Conclusions

The methodological framework proposed in this study demonstrated consistent results in capturing the temporal trends, spatial variability, and extreme behavior of the drought phenomenon in southern Portugal.

The findings provide a comprehensive evaluation of drought dynamics in the region over the past four decades:

(i) A consistent increase in median and third quartile CDD values over the last 20 years
(ii) The Joint Drought Index analysis highlights the importance of combining indicators to generate more comprehensive and robust insights into drought conditions. The increase in this index indicates a persistent intensification of the drought phenomenon.
(iii) The rise in the interquartile range (IQR) over the last two decades presents an additional concern: the increasing spatial uncertainty of drought, particularly in severe events, making the phenomenon locally more unpredictable.
(iv) The early warning alert system, based on machine learning short-term predictions of drought severity, proves to be a promising tool for mitigating the impacts of such events.

Finally, this study highlights the urgency of expanding this workflow to gain a deeper understanding of drought dynamics across the entire Mediterranean Basin—one of the most severely affected regions in the world.

Bibliography

1. Borg, I., Groenen, P.: Modern Multidimensional Scaling: Theory and Applications. Springer, New York (1997)
2. Costa, A.C., Soares, A.: Trends in extreme precipitation indices derived from a daily rainfall database for the South of Portugal. Int. J. Climatol. **29**(13), 1956–1975 (2009)
3. Cox, T.F., Cox, M.A.A.: Multidimensional Scaling. Chapman and Hall, London (1994)
4. De Iaco, S., Myers, D.E., Posa, D.: Total air pollution and space-time modelling. In: Monestiez, P., Allard, D., Froidevaux, R. (eds.) geoENV III - Geostatistics for Environmental Applications, Quantitative Geology and Geostatistics, vol. 11. Springer, Dordrecht (2001)
5. Deutsch, C.V., Journel, A.G.: GSLIB: Geostatistical Software Library and User's Guide, 2 edn. Oxford University Press, New York (1997)
6. Durao, R.M., Pereira, M.J., Costa, A.C., Delgado, J., Del Barrio, G., Soares, A.: Spatial-temporal dynamics of precipitation extremes in southern Portugal: a geostatistical assessment study. Int. J. Climatol. **30**(10), 1526–1537 (2010)
7. Gomes, M., Meira, A., Durao, R., Soares, A.: Geostatistical characterization of climate extremes dynamics: the drought phenomenon. In: Book of Abstracts - 12th International Geostatistics Congress (2024)
8. Hollister, B.E., Pang, A.: Interpolation of non-Gaussian probability distributions for ensemble visualization. In: Proceedings of the IEEE Visualization Posters (2013)
9. Horta, A., Soares, A.: Direct sequential co-simulation with joint probability distributions. Math. Geosci. **42**(3), 269–292 (2010)

10. Neves, J., Araujo, C., Soares, A.: Uncertainty integration in dynamic mining reserves. Math. Geosci. **53**(4), 737–755 (2021)
11. Oliveira, A.R., Branquinho, C., Pereira, M., Soares, A.: Stochastic simulation model for the spatial characterization of lung cancer mortality risk and study of environmental factors. Math. Geosci. **45**(4), 437–452 (2013)
12. Pedregosa, F., Varoquaux, G., Gramfort, A., Michel, V., Thirion, B., Grisel, O., Blondel, M., Prettenhofer, P., Weiss, R., Dubourg, V., Vanderplas, J., Passos, A., Cournapeau, D., Brucher, M., Perrot, M., Duchesnay, E.: Scikit-learn: machine learning in python. J. Mach. Learn. Res. **12**, 2825–2830 (2011)
13. Read, A.L.: Linear interpolation of histograms. Nucl. Instrum. Methods Phys. Res., Sect. A **425**(1–2), 357–360 (1999)
14. Rouhani, S., Wackernagel, H.: Multivariate geostatistical approach to space-time data analysis. Water Resour. Res. **26**(4), 585–591 (1990)
15. Ryan, C., Curley, M., Walsh, S., Murphy, C.: Long-term trends in extreme precipitation indices in Ireland. Int. J. Climatol. **42**(7), 4040–4061 (2022)
16. Scheidt, C., Caers, J.: Representing spatial uncertainty using distances and kernels. Math. Geosci. **41**(4), 397–419 (2009)
17. Soares, A.: Direct sequential simulation and cosimulation. Math. Geol. **33**(8), 911–926 (2001)
18. Sun, Y., Wang, Y., Zhang, M., Zeng, Z.: Summer extreme consecutive dry days over Northeast China in the changing climate: observed features and projected future changes based on CESM-LE. Front. Earth Sci. **11**, 1138985 (2023)
19. Todaro, V., D'Oria, M., Secci, D., Zanini, A., Tanda, M.G.: Climate change over the Mediterranean region: local temperature and precipitation variations at five pilot sites. Water **14**(16), 2499 (2022)
20. Torgerson, W.S.: Multidimensional scaling: I. Theory and method. Psychometrika **17**(4), 401–419 (1952)
21. Zhang, X., Alexander, L., Hegerl, G.C., Jones, P., Tank, A.K., Peterson, T.C., Trewin, B., Zwiers, F.W.: Indices for monitoring changes in extremes based on daily temperature and precipitation data. Wiley Interdiscip. Rev. Clim. Chang. **2**(6), 851–870 (2011)

Using Machine Learning Methods to Explore the Effects of Environmental Variables on Biodiversity

Iman Masoumi, Amirhossein Najafabadipour, Sabrina Maggio, and Giuseppina Giungato

Abstract Environmental factors, notably air pollutants and meteorological conditions, significantly impact biodiversity distribution and ecosystem dynamics. Among these, PM_{10} is identified as a major stressor, affecting plant physiological responses, reducing species abundance, and modifying soil microbial activity and nutrient cycling. These effects are often compounded when PM_{10} interacts with climatic conditions.

A comprehensive understanding of these combined effects is crucial for regions like Apulia, where biodiversity is shaped by both climate variability and localized human activities. The region remains underexplored in biodiversity modeling, particularly regarding the use of machine learning methods with the Multilevel Biodiversity Index (MBI).

This study aims to address this gap by a comparison of the performance of several Machine Learning (ML) models, including Random Forests (RFs), Support Vector Machines (SVMs), Extreme Gradient Boosting (XGBoost), Decision Trees (DTs), and Linear Regression (LR), for predicting MBI based on PM_{10} concentrations and meteorological variables. Among the models tested, DTs and XGBoost performed best, reliably predicting MBI based on various evaluation metrics.

Keywords Multilevel Biodiversity Index (MBI) · Machine learning (ML) · Fine particulate matter (PM_{10}) · Supervised machine learning algorithms

---

I. Masoumi
National Biodiversity Future Center, Palermo, Italy

A. Najafabadipour
Faculty of Mining Engineering, University of Jiroft, Jiroft, Iran

S. Maggio (✉) · G. Giungato
Department of Economic Sciences, University of Salento, Lecce, Italy
e-mail: sabrina.maggio@unisalento.it

S. De Iaco et al. (eds.), *Exploration of Spatio-Temporal Environmental Conditions: Harmonized Databases and Analytical Techniques*, Springer Proceedings in Mathematics & Statistics 531, https://doi.org/10.1007/978-3-032-17526-7_11

1 Introduction

Biodiversity is crucial for several essential ecological functions, including pollination, water filtration, and climate regulation. However, global biodiversity is currently undergoing a rapid and sustained decline, often referred to as the sixth mass extinction [8, 27, 29]. Localized pressures are responsible for the increase in this decline, especially in biodiversity hotspots such as Italy's Apulia region, which balances significant ecological value with notable vulnerability to both regional and global environmental drivers [4].

Indeed, Apulia hosts diverse ecosystems, including farmlands, forests, urban areas, and protected zones, which are fundamental for unique species and ecological integrity. Despite this richness, Apulia faces increasing threats from pollution, land use changes, and climate variability. Human-driven factors such as urban expansion, intensive agriculture, and diffuse water contamination contribute to ecosystem fragmentation, placing rare and endemic species under additional stress.

Biodiversity decline has broader implications for global sustainability targets, particularly concerning the United Nations Sustainable Development Goals (SDGs), such as Goal 13 *Climate Action* and Goal 15 *Life on Land*. Changes in climate, including rising temperatures, altered precipitation regimes, and increased frequency of extreme weather events, can accelerate biodiversity loss by disrupting ecological stability. Local environmental factors, including land degradation and high levels of air pollutants, like PM_{10}, produce a negative impact on ecosystem resilience.

Abiotic factors like air pollutants and meteorological conditions significantly influence biodiversity patterns. PM_{10} impacts plant processes, species richness, and soil function [24]. These effects are amplified by climate factors, with temperature and precipitation changes affecting habitats and productivity and extreme weather potentially increasing pollutant deposition [24, 26, 29]. In this context, an integrated analysis of multiple drivers is necessary for managing biodiversity in ecologically diverse regions like Apulia [24, 29].

Despite the strong capabilities of machine learning (ML) models like Support Vector Machines (SVMs), Extreme Gradient Boosting (XGBoost), and Decision Tree (DTs) in ecological modeling, limitations persist concerning data quality, spatial heterogeneity, and generalization [9, 15, 25, 32, 35–37, 42]. While remote sensing and citizen science initiatives have improved data coverage, integrating diverse sources into consistent predictive models remains challenging, particularly for citizen science data with uneven spatial effort [28]. Given these limitations, there is a clear need for unified frameworks that can integrate various data types and spatial dimensions, especially in data-constrained regions such as Apulia.

Recent research has shown the power of advanced ML techniques, including ensemble methods like stacking and bagging, to handle complex and nonlinear ecological datasets and enhance predictive accuracy, although these techniques are rarely applied for biodiversity prediction purposes in the Mediterranean area [1, 20], compared to other environmental contexts and sustainability areas [17–19].

This study aims to predict the Multilevel Biodiversity Index (MBI), a composite indicator of species richness and land cover diversity [10] in Apulia region. In particular, a comparison of the performance of four supervised machine learning algorithms such as SVMs, XGBoost, DTs, and LR is discussed. Spatially interpolated environmental variables (PM_{10}, atmospheric temperature, atmospheric pressure, and precipitation) are considered to evaluate model accuracy and identify the best approach for catching spatial variation in biodiversity patterns throughout the Apulia region.

The developed ML framework provides accurate, spatially explicit biodiversity forecasting, thereby fostering scientific comprehension and supporting strategic policy in the study area.

Furthermore, this research improves MBI modeling framework in Mediterranean and data-scarce landscapes by resolving existing methodological issues and increasing predictive reliability through comparative model evaluation.

This chapter is organized into three main sections. Section 2 details the methodology and data covering the study area, data acquisition, the creation of the Multilevel Biodiversity Index (MBI), as well as the machine learning methods used, along with their evaluation metrics. Section 3 presents the biodiversity modeling outcomes, specifically the spatial interpolation of environmental variables in Apulia and the prediction performance of the MBI. Finally, Sect. 4 summarizes the key findings and discusses their implications for biodiversity conservation.

2 Methodology and Data

The methodological framework for predicting the MBI in Apulia has involved several key stages: (1) data collection and preprocessing to correct spatial misalignments and address data quality issues, (2) computation of MBI and interpolation of environmental variables via Ordinary Kriging, (3) fitting of ML models (DTs, XGBoost, SVMs, and LR) to predict MBI, and (4) validation of results.

In particular, the first step has consisted of gathering essential spatial datasets for the Apulia region, including species occurrence records, land use and land cover (LULC) data, and environmental variables such as PM_{10} concentrations, temperature, atmospheric pressure, and precipitation. Then, all datasets have been preprocessed to correct spatial misalignments and address data quality issues.

Following this, meteorological and air quality variables have been interpolated using geostatistical methods implemented in dedicated routines of the GSLib software. This process has allowed for the creation of a uniform prediction grid across the study area, enabling consistent spatial integration of heterogeneous inputs suitable for ML applications. Subsequently, the modeling phase has been based on the comparison of four supervised ML algorithms: XGBoost, SVM, DT, and LR. The dataset has been split into 70% for training and 30% for testing to thoroughly evaluate model performance. Each model has been trained to estimate MBI based on selected environmental predictors. Finally, model performance has been assessed using various metrics, allowing comprehensive comparative evaluation of predictive accuracy.

2.1 Study Area and Data Acquisition

Apulia, located in Southern Italy, is a pivotal Mediterranean biodiversity hotspot featuring diverse ecosystems like farmlands, forests, urban areas, and protected zones. It is a crossroads of species and hosts vital habitats such as temporary ponds and coastal lagoons, essential for unique species and ecological integrity [7, 10]. However, the region faces growing threats from pollution, land use changes, and climate variability [30]. Human activities, including urban expansion, intensive agriculture, and water contamination, cause ecosystem fragmentation, endangering rare and endemic species. This highlights the need for targeted regional evaluations to comprehend and mitigate biodiversity loss in Apulia's vulnerable environments. For this aim, the analysis has started by gathering the following spatial datasets for the Apulia region:

- *Species Occurrence Data*, obtained from the Global Biodiversity Information Facility (GBIF) (https://www.gbif.org/). The dataset has comprised 729 species records from 1,014 georeferenced locations. These records have spanned three taxonomic kingdoms: Animalia (519 species), Plantae (206 species), and Fungi (4 species). Preprocessing for reliability, consistent with GBIF and Darwin Core standards, has been performed, which has included the removal of duplicate records, correction of geospatial inaccuracies, and exclusion of fossil data, enhancing data usability for MBI modeling.
- *Land Use and Land Cover (LULC) Data*, obtained from Italy's National Geoportal (http://www.pcn.minambiente.it/mattm/servizio-wms/). This detailed dataset, adhering to CORINE Land Cover standards, has been essential for analyzing biodiversity distribution and mapping key habitats like Mediterranean Temporary Ponds and forests.
- *Environmental Variables* (including PM_{10} concentrations, temperature, atmospheric pressure, and precipitation), collected from ARPA Puglia (Regional Agency for Environmental Protection) monitoring networks (https://www.arpa.puglia.it). These data have been collected from a range of monitoring station types, including traffic, industrial, and background sites, located in major urban centers such as Foggia, Bari, Brindisi, Taranto, and Lecce. PM_{10} is known to pose significant anthropogenic pressure, impacting plants and soil. Exposure to fine particulate matter ($PM_{2.5}$ and PM_{10}) negatively affects stomatal function and physiological processes in plants, while also promoting soil acidification, which disrupts microbial communities and nutrient cycling in Mediterranean ecosystems. In Mediterranean regions like Apulia, PM_{10} levels, often affected by African dust intrusions, can significantly influence both soil characteristics and ecosystem functioning. Temperature variations, including long-term warming

trends, contribute to shifts in species distributions, reduced habitat suitability, and increased mortality during extreme heat periods. Elevated temperatures within Mediterranean ecosystems can also affect phenological timing, decrease pollination efficiency, and disrupt plant communities. Biodiversity is influenced by atmospheric pressure through its effects on microclimates, soil moisture retention, and nutrient cycling, particularly in coastal and mountainous areas. The distribution of precipitation is crucial for governing water availability and ecosystem productivity. Reduced rainfall exacerbates drought conditions, negatively affecting soil microbial activity, plant biomass, and water-dependent species.

2.2 *Multilevel Biodiversity Index (MBI) and Spatial Interpolation*

To evaluate spatial biodiversity patterns, the Multilevel Biodiversity Index (MBI) has been computed as an integrated metric combining species richness with land cover diversity, as follows:

$$\mathrm{MBI} = (\mathrm{S} \times \mathrm{LULC})/\mathrm{A}$$

where:

- *S* represents true species occurrence, reflecting the effective number of existing species.
- *LULC* denotes the number of land use/land cover diversity classes.
- *A* is the sampled area in (km^2).

This index standardizes biodiversity values across a spatial grid, making it possible to compare ecologically heterogeneous landscapes. Higher MBI values indicate greater ecological diversity, while lower values suggest limited biodiversity or uniform land cover types. The MBI values were derived from species occurrence data aggregated on a regular grid with dimensions of 15 × 13 and cell sizes of 19.5 km × 19.5 km. The results have shown elevated MBI values in coastal areas, where species richness and landscape diversity are higher, and reduced values in inland zones characterized by intensive agriculture and industrial activity. Unlike traditional indices, MBI has provided a more holistic view of ecological complexity by integrating land cover diversity and rarity, which has proven especially useful in regions with high environmental variability like Apulia.

For the meteorological and air quality variables, Ordinary Kriging has been used for spatial interpolation. This geostatistical method has been chosen for its ability to account for spatial autocorrelation and produce unbiased, reliable predictions. This process created a uniform prediction grid (19.5 km × 19.5 km) across the study area, ensuring consistent spatial integration of heterogeneous inputs suitable for

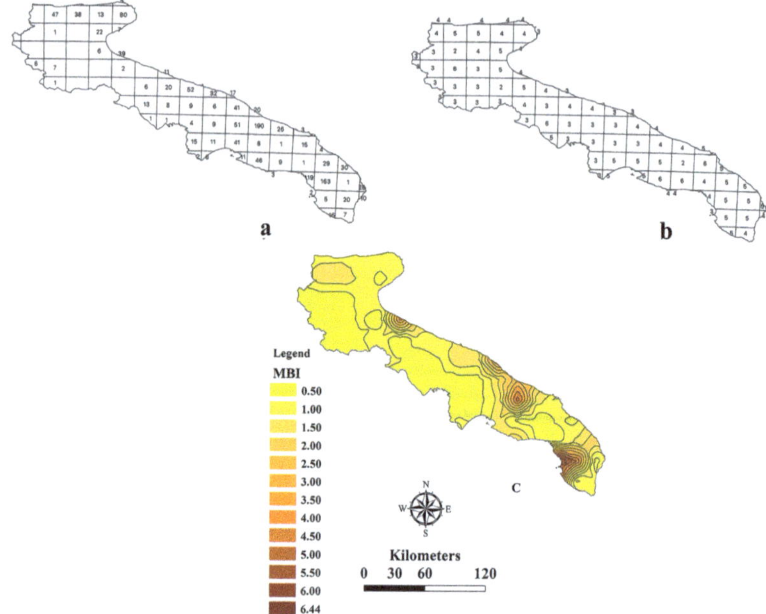

Fig. 1 (**a**) Number of species occurrence, (**b**) number of LULC classes, and (**c**) MBI contour map in the Apulia region

ML applications. Structural analysis, involving variogram estimation and model selection, has been conducted for each variable under study to guarantee spatial continuity and compatibility with the biodiversity modeling.

Figure 1 illustrates the regional spatial patterns of species occurrences, LULC diversity, and the derived MBI.

2.3 *Machine Learning Methods*

Machine learning methods have increasingly been used in biodiversity studies due to their ability to manage complex ecological datasets, identify nonlinear relationships, and select influential environmental predictors [6, 12]. In this study, four supervised ML algorithms (XGBoost, DT, SVM, and LR) have been applied to model MBI in the Apulia region.

XGBoost is a supervised machine learning algorithm that uses an additive process called gradient boosting to build predictive models from an ensemble of decision trees. It excels at identifying complex, nonlinear relationships and is both scalable and robust to missing data. However, its performance relies heavily on hyperparameter tuning [32], and it can be computationally expensive for very large datasets [12]. The model has been successfully used in a variety of ecological contexts, such as modeling forest biomass and predicting species richness.

The Decision Tree (DT) algorithm is a nonparametric supervised model that recursively partition data for predictions, valued for interpretability and handling nonlinear interactions [13]. However, DT is prone to overfitting if the tree becomes too complex, which can be mitigated through pruning [22]. In ecology, DTs have been applied to various tasks, from predicting vegetation types to analyzing bird migration.

Support Vector Machine (SVM) is a robust supervised method for classification and regression, effective with high-dimensional, nonlinear data. SVM identifies optimal separating hyperplanes using support vectors and kernel functions [14]. It also has a strong ability to prevent overfitting [43]; however, SVM can be computationally expensive with large datasets and sensitive to parameter tuning. In ecological studies, SVM has been successfully applied to modeling forest diversity, assessing habitat suitability, and predicting species richness [5, 6, 37].

Linear Regression (LR) is a fundamental and efficient statistical method used to model the linear relationship between a dependent variable (like a biodiversity index) and one or more independent variables (like environmental factors) [23]. However, its primary limitation is its inability to effectively model nonlinear relationships and complex ecological patterns, which are very common in biodiversity data. Despite this limitation, LR remains a widely used tool for evaluating how land use impacts species richness or predicting biodiversity changes due to climate variation.

2.4 *Model Evaluation Metrics*

In this research, the forecasting capability of the machine learning models has been evaluated through the application of several widely used statistical measures. The metrics employed have encompassed the Nash-Sutcliffe Efficiency (NSE) [2, 31], the Mean Absolute Percentage Error ($MAPE$) [11, 16, 33, 39], the Root Mean Square Error ($RMSE$) [34], the coefficient of determination (R^2) [3], as well as the Williams plot diagnostics.

Nash-Sutcliffe Efficiency (NSE) measures the relative magnitude of residual variance compared to the variance of the observed data. Its values range from $-\infty$ to 1, with 1 denoting a perfect match between observed and predicted values. It is computed as follows:

$$NSE = 1 - \frac{\sum_{i=1}^{n}(O_i - P_i)^2}{\sum_{i=1}^{n}(O_i - \bar{O})^2}, \tag{1}$$

where O_i and P_i are observed and predicted values, respectively, n is the number of observations, and $\bar{O}$ is the mean of observed values.

Mean Absolute Percentage Error ($MAPE$) determines the average magnitude of a model's prediction errors as a percentage of the observed values. This provides a clear and consistent way to interpret a model's precision, even when comparing forecasts for different metrics. It is defined as follows:

$$MAPE = \frac{1}{n}\sum_{i=1}^{n}\left|\frac{O_i - P_i}{O_i}\right| \cdot 100, \tag{2}$$

where n is the number of observations.

Root Mean Square Error ($RMSE$) is a common metric used to measure the average difference between a model's predicted values and the actual observed values. It is often used to quantify a model's accuracy, with lower values indicating a better fit. It is given by

$$RMSE = \sqrt{\frac{1}{n}\sum_{i=1}^{n}(O_i - P_i)^2}. \tag{3}$$

Coefficient of determination (R^2) is a statistical measure that evaluates the proportion of variance in observed data explained by the model. This metric is mathematically similar to the NSE. While both are powerful tools for evaluating model performance, they are used in different contexts. A key distinction is that while R^2 is typically used for assessing how well a linear model fits the original data, NSE is more commonly used to evaluate the predictive power of a model on new or simulated data, especially in fields like hydrology. It is expressed as

$$R^2 = 1 - \frac{\sum_{i=1}^{n}(O_i - P_i)^2}{\sum_{i=1}^{n}(O_i - \bar{O})^2}. \tag{4}$$

R^2 values range from 0 to 1, with higher values indicating better model performance and explanatory power.

Williams plot diagnostics employ a combination of leverage and standardized residuals to evaluate a model's validity and identify outliers. The values for leverage (h_i) and standardized residuals (r_i) are computed as follows:

$$\begin{aligned} h_i &= x_i^T (X^T X)^{-1} x_i, \\ r_i &= \frac{O_i - P_i}{\sigma}, \end{aligned}$$

where x_i^T represents the predictor vector for observation i, X is the matrix of predictors, and σ denotes the standard deviation of residuals.

Collectively, these diagnostic tools have offered a complete appraisal of the model's predictive performance, the error magnitude, and its overall reliability.

3 Environmental Variables and Biodiversity Prediction

This section details the spatial interpolation results for environmental variables and the MBI's predictive performance.

3.1 Geostatistical Interpolation of Environmental Indicators

The spatial mapping of several key environmental factors across the Apulia region has been conducted. This has been achieved using Ordinary Kriging, a geostatistical technique widely recognized for its ability to address spatial continuity and provide unbiased estimations in a variety of environmental contexts [38, 40]. In particular, the structural analysis, characterized by the variogram estimation and the model selection, has firstly been carried out for the environmental indicators. According to the spatial modeling results and to ensure spatial consistency with the MBI dataset, Kriging interpolation has been performed on a uniform 19.5 km × 19.5 km grid.

The resulting interpolated maps (Fig. 2) have highlighted the distinct spatial diversity within the region and have also facilitated the evaluation of environmental drivers that have influenced biodiversity characteristics throughout Apulia.

The generated colormaps are used as inputs for biodiversity modeling as they capture the landscape's essential abiotic patterns. These interpolated outputs provide a continuous spatial representation of key environmental variables (like PM_{10}, atmospheric temperature, atmospheric pressure, and precipitation), which effectively show the region's environmental variability. By applying a uniform grid, the datasets create a consistent foundation for machine learning applications, ensuring all environmental predictors are aligned within a coherent spatial framework. As

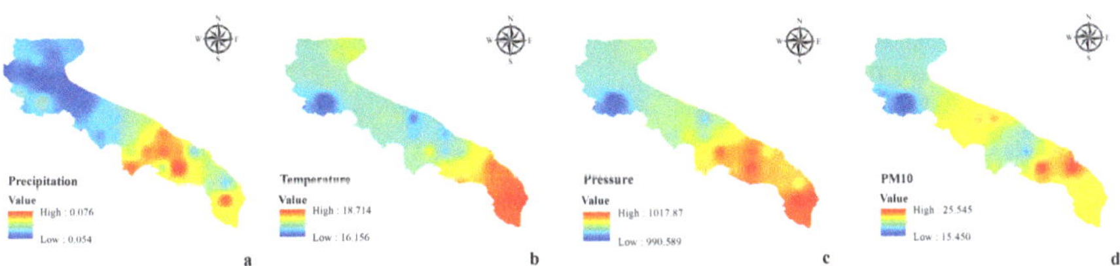

Fig. 2 Colormaps illustrating (**a**) Precipitation, (**b**) Temperature, (**c**) Atmospheric Pressure, and (**d**) PM_{10} concentrations in Apulia, obtained by applying Ordinary Kriging

shown in Fig. 2, these maps highlight spatial differences and are fundamental to evaluate the influence of environmental factors on biodiversity patterns in Apulia.

Elevated PM_{10} concentrations are found in urban and industrial hubs (e.g., Taranto, Brindisi, and Bari). These areas are correlated with diminished MBI values, suggesting a potential link between atmospheric pollution and ecological stress. Temperature shows both latitudinal and coastal-to-inland variations, with warmer conditions prevalent in the southeastern peninsula and coastal zones, aligning with established Mediterranean climate tendencies and species range shifts. Atmospheric pressure tends to be lower over elevated terrains and more stable across coastal lowlands, reflecting the influence of topography on atmospheric mass. Precipitation exhibits a north-to-south gradient, with higher amounts in the northern Apennine foothills and reduced rainfall in the drier southeastern plains, a critical factor for water-dependent species and ecosystems. These interpolated maps define abiotic gradients directly informing biodiversity modeling. Indeed, by comparing the interpolation results with MBI data, it has been highlighted that regions characterized by elevated particulate matter levels, scarce precipitation, or heightened temperatures have generally displayed reduced biodiversity, underscoring the predictive significance of these factors in spatial ecological risk assessments.

3.2 Model Accuracy and Validation for MBI Prediction

The forecasting accuracy of four machine learning algorithms (XGBoost, DT, SVM, and LR) for MBI estimation in Apulia has been assessed using a multi-metric validation methodology. These models have been trained with fine particulate matter levels, temperature, atmospheric pressure, and precipitation as input features. Model efficacy has been gauged using the NSE and MAPE, complemented by diagnostic tools such as scatterplots comparing predicted versus actual values, cumulative distribution functions of RMSE, Williams plots, and a Taylor diagram [41].

With the highest predictive accuracy (NSE: 0.87, MAPE: 1.08%), the Decision Tree (DT) model has demonstrated strong concordance between observed and predicted MBI values.

XGBoost has also produced commendable performance (NSE: 0.81, MAPE: 3.71%). Conversely, SVM (NSE: 0.54, MAPE: 4.58%) and LR (NSE: 0.16, MAPE: 6.51%) models have exhibited inferior predictive capabilities, indicative of their limited effectiveness in capturing the intricate spatial and nonlinear characteristics of Mediterranean biodiversity.

Williams plot diagnostics for all models have indicated that the majority of standardized residuals fell within the ± 3 range ($\pm$ 3 times the standard deviation) and within the leverage boundaries, suggesting reliable and generalizable predictions. Outliers might reflect ecotonal regions (e.g., transitions between urban and rural areas or coastal buffer zones) where ecological variability complicates prediction accuracy. A Taylor diagram comparison has shown the DT model achieved the highest correlation coefficient (above 0.95) and a standard deviation closely aligned with observed data, while XGBoost also performed strongly. SVM and LR models were positioned farther from the reference point, indicating lower predictive accuracy.

Finally, relevancy factor analysis has quantified the contribution of each input variable to the model predictions [21]. Spatial coordinates were identified as the most influential predictors, suggesting the significant influence of underlying spatial gradients and latent spatial heterogeneity not entirely explained by environmental variables alone. Among the atmospheric inputs, pressure and temperature have shown the strongest impact, followed by precipitation and PM_{10}.

4 Conclusions

This study highlights the crucial role of algorithm selection in ecological modeling, particularly for MBI prediction over the Apulia region. A comparative analysis of four supervised machine learning models (XGBoost, DT, SVM, and LR) consistently demonstrated that tree-based methods, specifically Decision Tree (DT) and XGBoost, outperformed linear and kernel-based models across all validation metrics.

The DT model proved to be the most reliable and accurate, exhibiting excellent concordance between observed and predicted MBI values (NSE: 0.87 and MAPE: 1.08%). Its strong performance, supported by diagnostic plots, can be attributed to its capacity to effectively model the complex, nonlinear interactions vital for spatially heterogeneous ecosystems like those found in Apulia. EXGBoost also delivered a strong performance (NSE: 0.81 and MAPE: 3.71%). Conversely, the SVM and LR models showed limited predictive capability, struggling to handle the intricate spatial and environmental variability characteristic of Mediterranean landscapes.

From a conservation standpoint, the accuracy of DT and XGBoost in producing spatially explicit biodiversity maps is paramount for informing adaptive management strategies in sensitive areas. Their ability to integrate complex abiotic gradients, including PM_{10} concentrations and precipitation, highlights their utility in addressing real-world ecological challenges, as these factors are known to influence biodiversity.

In conclusion, this research confirms that carefully selected and validated machine learning methods, particularly tree-based ones, offer powerful tools for MBI prediction and provide a transferable framework for biodiversity modeling in other Mediterranean and data-scarce regions. Future investigations should broaden this approach by incorporating additional environmental descriptors (e.g., vegetation indices, land use shifts, and soil characteristics) and exploring hybrid modeling architectures or temporal data integration to refine predictive accuracy and ecological insight, thereby supporting targeted, evidence-based conservation strategies.

Bibliography

1. Abid, M., González, J.A., de Rivera, Ó.R., Moraga, P.: Mapping the spatio-temporal distribution of burned areas in the Amazon from 2001 to 2020: an ensemble modeling approach. Environ. Ecol. Stat. **32**(2), 707–734 (2025)
2. Ahmad, A., Sulaiman, M., Alhindi, A., Aljohani, A.J.: Analysis of temperature profiles in longitudinal fin designs by a novel neuroevolutionary approach. IEEE Access **8**, 113285–113308 (2020)
3. Al-Shboul, K.F., Almasabha, G., Shehadeh, A., Alshboul, O.: Exploring the efficacy of machine learning models for predicting soil radon exhalation rates. Stochastic Environ. Res. Risk Assess. **37**, 4307–4321 (2023)
4. Antonelli, A., Dhanjal-Adams, K.L., Silvestro, D.: Integrating machine learning, remote sensing and citizen science to create an early warning system for biodiversity. Plants People Planet **5**, 307–316 (2023)
5. Asadi, H., Jalilvand, H., Tafazoli, M., Hosseini, S.F.: Modeling habitat suitability of Quercus castaneifolia in the Hyrcanian forest: a comprehensive integration of environmental factors for conservation insights. Biodivers. Conserv. **34**(1), 315–334 (2025)
6. Bayat, M., Burkhart, H., Namiranian, M., Hamidi, S.K., Heidari, S., Hassani, M.: Assessing biotic and abiotic effects on biodiversity index using machine learning. Forests **12**(4), 461 (2021)
7. Beccarisi, L., Zuccarello, V., Accogli, R., Belmonte, G.: Biodiversity and possible bioindicators of Mediterranean temporary ponds in southern Apulia, Italy. Diversity **16**(9), 559 (2024)
8. Bellard, C., Bertelsmeier, C., Leadley, P., Thuiller, W., Courchamp, F.: Impacts of climate change on the future of biodiversity. Ecol. Lett. **15**, 365–377 (2012)
9. Cai, R., Xie, S., Wang, B., Yang, R., Xu, D., He, Y.: Wind speed forecasting based on Extreme Gradient Boosting. IEEE Access **8**, 175063–175069 (2020)
10. Cazzolla Gatti, R., Notarnicola, C.: A novel multilevel biodiversity index (MBI) for combined field and satellite imagery surveys. Global Ecol. Conserv. **13**, e00361 (2018)
11. Chaibi, M., Ben Ghoulam, E.M., Khallouk, N., Tarik, L., El Yousfi, Y., El Hmaidi, A., Berrada, M., Mabrouki, J.: A novel fuzzy-multi-criteria-GIS-machine learning approach for onshore wind power plant site selection. Euro-Mediterr. J. Environ. Integr. **10**, 1025–1045 (2025)
12. Chang, G.J.: Biodiversity estimation by environment drivers using machine/deep learning for ecological management. Eco. Inform. **78**(10), 102319 (2023)
13. Coops, N.C., Waring, R.H., Beier, C., Roy-Jauvin, R., Wang, T.: Modeling the occurrence of 15 coniferous tree species throughout the Pacific Northwest of North America using a hybrid approach of a generic process-based growth model and decision tree analysis. Appl. Veg. Sci. **14**(3), 402–414 (2011)
14. Cortes, C., Vapnik, V.: Support-vector networks. Mach. Learn. **20**, 273–297 (1995)
15. De Iaco, S., Hristopulos, D.T., Lin, G.: Special issue: geostatistics and machine learning. Math. Geosci. **54**(3), 459–465 (2022)
16. de Myttenaere, A., Golden, B., Le Grand, B., Rossi, F.: Mean absolute percentage error for regression models. Neurocomputing **192**, 38–48 (2016)
17. Distefano, V., Palma, M., De Iaco, S.: Multi-class random forest model to classify wastewater treatment imbalanced data. Soc.-Econ. Plan. Sci. **95**, 102021 (2024)
18. Distefano, V., De Iaco, S., Masoumi, I.: Artificial neural network optimization to estimate radon in soil. Stat. Anal. Data Mining ASA Data Sci. J. **18**(5), e70036 (2025)
19. Distefano, V., Gentile, V., Cucurachi, P.A., De Iaco, S.: Sustainable finance disclosure versus performance: a clustering approach. Bus. Strateg. Environ. **34**, 7837–7850 (2025)
20. Farzipour, A., Elmi, R., Nasiri, H.: Detection of monkeypox cases based on symptoms using XGBoost and Shapley Additive Explanations Methods. Diagnostics **13**(14), 2391 (2023)

21. Fathinasab, M., Shahbazi, K.: A new correlation for estimation of minimum miscibility pressure (MMP) during hydrocarbon gas injection, J. Pet. Explor. Prod. Technol. **10**(6), 2349–2356 (2020)
22. Felix, F.C., Kratz, D., Ribeiro, R., Nogueira, A.C.: Machine learning in the identification of native species from seed image analysis. J. Seed Sci. **46**(1), e202446002 (2024)
23. Gómez-Tolosa, M., González-Soriano, E., Mendoza-Cuenca, L.F., Pérez-Mungúıa, R.M., Rioja-Paradela, T.M., Espinoza-Medinilla, E.E., Ortega-Salas, H., Rivera-Velázquez, G., Penagos-Garćıa, F.E., López, S.: The use of highly diverse clades as a surrogate for habitat integrity analysis: Argia damselflies as a practical tool for rapid assessments. Environ. Sci. Pollut. Res. **29**, 24334–24347 (2022)
24. Grantz, D.A., Garner, J.H.B., Johnson, D.W.: Ecological effects of particulate matter. Environ. Int. **29**(2–3), 213–239 (2003)
25. Grillo, M., Schiaparelli, S., Durazzano, T., Guglielmo, L., Granata, A., Huettmann, F.: Machine learning applied to species occurrence and interactions: the missing link in biodiversity assessment and modelling of Antarctic plankton distribution. Ecol. Proces. **13**(1), 56 (2024)
26. Habibullah, M.S., Din, B.H., Tan, S.H., Zahid, H.: Impact of climate change on biodiversity loss: global evidence. Environ. Sci. Poll. Res. **29**, 1073–1086 (2022)
27. López Rodŕıguez, S., van Bussel, L.G.J., Alkemade, R.: Classification of agricultural land management systems for global modeling of biodiversity and ecosystem services. Agric. Ecosyst. Environ. **360**, 108795 (2024)
28. Luiselli, L., Pacini, N.: Remote sensing that makes sense in ecological research—from pixels to conservation. Afr. J. Ecol. **63**(1), e70002 (2025)
29. Mantyka-pringle, C.S., Martin, T.G., Rhodes, J.R.: Interactions between climate and habitat loss effects on biodiversity: a systematic review and meta-analysis. Glob. Chang. Biol. **18**(4), 1239–1252 (2012)
30. Masoumi, I., De Iaco, S., Maggio, S.: Exploring land use and land cover changes in Apulia, Italy: random forest approach utilizing remote sensing data. In: Pollice, A., Mariani, P. (eds.) Methodological and Applied Statistics and Demography I. SIS 2024, pp. 259–264. Italian Statistical Society Series on Advances in Statistics. Springer, Cham (2025). https://doi.org/10.1007/978-3-031-64346-0_44
31. McCuen, R.H., Knight, Z., Cutter, A.G.: Evaluation of the Nash–Sutcliffe efficiency index. J. Hydraul. Eng. **11**(6), 597–602 (2006)
32. Militino, A.F., Goyena, H., Pérez-Goya, U., Ugarte, M.D.: Logistic regression versus XGBoost for detecting burned areas using satellite images. Environ. Ecol. Stat. **31**, 57–77 (2024)
33. Nabillah, I., Ranggadara, I.: Mean Absolute Percentage Error untuk Evaluasi Hasil Prediksi Komoditas Laut. J. Inf. Syst. **5**(2), 250–255 (2020)
34. Najafabadipour, A., Kamali, G.R., Nezamabadi-pour, H.: The innovative combination of time series analysis methods for the forecasting of groundwater fluctuations. Water Resour. **49**(2), 283–291 (2022)
35. Nel, H., Mishra, A.K., Schonken, F.: EcoSonicML: Harnessing machine learning for biodiversity monitoring in South African Wetlands. SN Comput. Sci. **5**(5), 513 (2024)
36. Novi, L., Bracco, A.: Machine learning prediction of connectivity, biodiversity and resilience in the coral triangle. Commun. Biol. **5**(1), 1359 (2022)
37. Park, D.S., Willis, C.G., Xi, Z., Kartesz, J.T., Davis, C.C., Worthington, S.: Machine learning predicts large scale declines in native plant phylogenetic diversity. New Phytol. **227**, 1544–1556 (2020)
38. Pellicone, G., Caloiero, T., Modica, G., Guagliardi, I.: Application of several spatial interpolation techniques to monthly rainfall data in the Calabria region (Southern Italy). Int. J. Climatol. **38**(9), 3651–3666 (2018)
39. Prayudani, S., Hizriadi, A., Lase, Y.Y., Fatmi, Y., Al-Khowarizmi, A.: Analysis accuracy of forecasting measurement technique on random k-nearest neighbor (RKNN) using MAPE and MSE. J. Phys. Conf. Ser. **1361**(1), 012089 (2019)
40. Susanto, F., De Souza, P., He, J.: Spatiotemporal interpolation for environmental modelling. Sensors **16**(8), 1245 (2016)

41. Taylor, K.E.: Summarizing multiple aspects of model performance in a single diagram. J. Geophys. Res. Atmos. **106**(D7), 7183–7192 (2001)
42. Tian, H., Huang, L., Hu, S., Wu, W.: A modified machine learning algorithm for multi-collinearity environmental data. Environ. Ecol. Stat. **31**(4), 1063–1083 (2024)
43. Wang, D., Wang, M., Qiao, X.: Support vector machines regression and modeling of greenhouse environment. Comput. Electron. Agric. **66**(1), 46–52 (2009)

Modeling and Prediction of Ground-Level Ozone Concentrations in a Spatiotemporal Multivariate Context

Claudia Cappello, Klaus Nordhausen, and Monica Palma

Abstract Accurate prediction of ground-level ozone concentrations is crucial for public health and environmental protection. Air quality monitoring processes require very high spatial resolution information for several hazardous air pollutants, as well as for those meteorological conditions that most influence air quality. The available datasets usually refer to long time series recorded irregularly over the area of interest or at a coarse spatial resolution. In this context, effective modeling and prediction methods are needed to forecast ozone levels in a high spatial resolution. In the chapter, the blind source separation-based approach is carried out to model and predict ground-level ozone concentrations in a multivariate framework.

Keywords Blind source separation model · Latent components · High-resolution prediction

1 Introduction

Modeling and predicting ground-level ozone concentrations pose significant challenges due to the strong spatial and temporal variability of atmospheric and emission-related processes. This secondary pollutant, formed through nonlinear photochemical reactions involving nitrogen oxides and volatile organic compounds under sunlight and stagnant meteorological conditions, represents a major envi-

C. Cappello
Department of Economic Sciences, University of Salento, Lecce, Italy

K. Nordhausen
Department of Mathematics and Statistics, University of Helsinki, Helsinki, Finland

M. Palma (✉)
Department of Economic Sciences, University of Salento, Lecce, Italy

National Centre for HPC, Big Data and Quantum Computing, Bologna, Italy
e-mail: monica.palma@unisalento.it

S. De Iaco et al. (eds.), *Exploration of Spatio-Temporal Environmental Conditions: Harmonized Databases and Analytical Techniques*, Springer Proceedings in Mathematics & Statistics 531, https://doi.org/10.1007/978-3-032-17526-7_12

ronmental and public health concern [36, 43]. An accurate modeling of such multivariate spatiotemporal data (MSTD) requires addressing complex dependencies both among variables and across spatial and temporal domains [17, 39, 41]. Many existing statistical approaches often neglect spatiotemporal dependencies or make restrictive assumptions on the process describing the data at hand, which can limit their applicability and lead to suboptimal outcomes. To simplify this demanding task, a common approach is to assume that observed variables are generated from fewer independent or uncorrelated latent components. Hence the recovered components are modeled univariately, reducing computational and conceptual complexity. In Multivariate Geostatistics, the aim of modeling the matrix-valued covariance function to explain direct and cross spatiotemporal dependence is reached by linearly combining univariate models associated with latent uncorrelated processes, identified through joint diagonalization of covariance matrices computed at distinct lags [6, 7, 18, 20, 23, 26, 27]. Moreover, vectorial data in two dimensions have found a reasonable characterization in the theory of complex-valued random fields, early introduced for 1D domains by [42] and for spatial domains in [19, 28, 29, 41], but some developments in parametric complex-domain covariance models appear in [11, 14, 37, 38], including advances in computational issues [9] and in a spatiotemporal-context [3, 4, 10, 12]. Another prominent approach, in the signal processing framework, is the Blind Source Separation (BSS) [1, 35]. BSS assumes that the observed multivariate data arise from linear combination of latent, independent processes and aims to estimate these hidden components and the corresponding unmixing matrix. This framework has been extended to the spatial [30, 33, 34] and spatiotemporal context [8, 31], as well as in a nonlinear framework [40]. By incorporating the spatiotemporal structures, ST-BSS enables the decomposition of complex spatiotemporal datasets into independent latent components.

In the present chapter, the ST-BSS modeling framework is applied to daily observations of ozone, nitrogen dioxide, nitric oxide, particulate matter, and meteorological variables collected across Northern Italy from 2013 to 2023. The proposed decomposition allows us to capture the main latent drivers of spatiotemporal variability and to produce high-resolution forecasts of ground-level ozone concentrations for unobserved time points. The results demonstrate that the ST-BSS based approach effectively models complex spatiotemporal dependencies and provides accurate ozone predictions, supporting both environmental assessment and air quality management strategies.

The chapter is structured as follows. After a brief review of the ST-BSS approach, highlighting its main features and flexibility in modeling an MSTD (Sect. 2), a case study on air pollutants and meteo-climatic variables is thoroughly discussed in the modeling and prediction frameworks (Sect. 3), then forecasts of the primary variable, the ground-level ozone concentration, are provided at very high spatial resolution (Sect. 4), and finally, some comments with discussion close the chapter (Sect. 5).

2 The ST-BSS Based Approach

ST-BSS is a statistical methodology employed to model and predict complex multivariate spatiotemporal data by simplifying the cross-variable dependencies. This approach derives from the signal processing community and has also been adopted for spatiotemporal data [31]. Its fundamental concept posits that observed data are a linear mixture of hidden (latent) processes. Hence, the primary objective of this approach is dimension reduction and simplification: by recovering these latent components, the complex multivariate spatiotemporal problem is reduced to modeling several univariate processes. Once modeled, the predicted latent components can be transformed back to the original observation space using the inverse of the unmixing matrix.

The ST-BSS model is formally expressed as $\mathbf{X}(\mathbf{s}, t) = \mathbf{A}\mathbf{Z}(\mathbf{s}, t) + \boldsymbol{\mu}$, where $\mathbf{X}(\mathbf{s}, t)$ denotes the p-variate observed random field, $\mathbf{Z}(\mathbf{s}, t)$ comprises p second-order stationary and independent components (ICs), $\mathbf{A}$ is a full-rank mixing matrix, and $\boldsymbol{\mu}$ is a deterministic location vector.

The latent field $\mathbf{Z}(\mathbf{s}, t)$ is assumed to have a zero expected value and unit variance, and its covariance matrix $\mathbf{D}(\mathbf{h}_s, h_t)$ is diagonal, with its diagonal elements representing the covariances of individual latent process components $c_l(\mathbf{h}_s, h_t)$. This indicates that the components are uncorrelated in both space and time. Hence, modeling p independent univariate processes is significantly more tractable than dealing with a comprehensive p-variate process [31]. The ST-BSS approach requires the identification of an unmixing matrix $\mathbf{W}$ that can recover $\mathbf{Z}(\mathbf{s}, t)$ from $\mathbf{X}(\mathbf{s}, t)$. This unmixing matrix $\mathbf{W}$ is derived through the joint diagonalization of scatter matrices that quantify spatiotemporal second-order dependence. These are typically Local Autocovariance Function (LACF) matrices, which are defined using specific spatiotemporal kernel functions such as Ball, Ring, or Gauss kernels. The stSOBI (spatiotemporal Second-Order Blind Identification) estimator is a commonly employed method, which maximizes a criterion based on diagonalizing multiple LACFs while constraining the covariance matrix of the unmixed data to be the identity matrix.

The ST-BSS workflow consists of three primary steps: 1. Estimate the unmixing matrix $\mathbf{W}$ and the mean $\boldsymbol{\mu}$ to derive the latent processes $\mathbf{Z}(\mathbf{s}, t)$. 2; Model each component of $\mathbf{Z}(\mathbf{s}, t)$ as a univariate process. 3; Back-transform the quantities of interest to the original scale using the inverse of the unmixing matrix, $\mathbf{W}^{-1}$ (and $\boldsymbol{\mu}$).

3 Spatiotemporal Modeling of Ozone Concentrations

Ground-level ozone (O3) arises from complex photochemical processes influenced by multiple interacting factors. The key determinants include (i) nitrogen oxides, mainly emitted from vehicular traffic and industrial activities, (ii) volatile organic compounds, originating from both anthropogenic and biogenic sources, (iii) particulate matter, which can affect photochemical reactions and atmospheric radiation balance, and (iv) meteorological conditions, such as air temperature, relative humidity, and solar radiation, which modulate the chemical reaction rates and pollutant dispersion. Accurate modeling and prediction of O3 concentrations require consideration of these correlated variables and their spatiotemporal dependence structures. In this context, the application of the ST-BSS approach and the corresponding independent latent components underlying the multivariate spatiotemporal process can be used to make predictions of the target variable in a multivariate framework. In particular, the case study focuses on Northern Italy and analyzes the joint behavior of air pollutants and meteorological variables recorded daily between 2018 and 2023 at a network of 143 monitoring stations. In particular, the following variables have been considered:

1. in situ ozone (O3) and nitrogen dioxide (NO2) concentrations ($\mu g/m^3$), as well as particulate matter with diameter below 10 μm (PM10)
2. satellite-derived data for nitric oxide (NO) concentrations ($\mu g/m^3$), air temperature (°C), relative humidity ($\%$), volatile organic compounds (VOC, g/m), solar radiation (W/m^2), and wind speed (m/s)

In situ data have been obtained from the European Air Quality Portal, while satellite products have been retrieved from the Copernicus Atmosphere Monitoring Service (CAMS).

As shown in Fig. 1, each variable exhibited a marked periodic component. Thus, it has been computed using daily averages and subsequently removed to ensure stationarity of the series. In addition, since the deseasonalized variables are characterized by different magnitudes, a standardization procedure has been applied, and scaled variables with zero mean and unit variance have been considered in the next stages of the analysis.

3.1 ST-BSS Modeling Steps

This section discusses the procedure adopted to identify the hidden components characterizing the investigated phenomenon using the ST-BSS approach and available in the R package `SpaceTimeBSS` [32]. The ST-BSS analysis follows the workflow described below:

- I. **Estimation of the Unmixing Matrix and Latent Processes**
  This step is crucial for separating the observed mixed signals into independent latent components. The stSOBI method has been chosen for this purpose. This approach is specifically designed for stationary spatiotemporal data and performs source recovery by jointly diagonalizing the covariance matrix and several LACF matrices.

  To capture the spatiotemporal dependencies required by stSOBI, 15 spatiotemporal kernel functions have been selected:

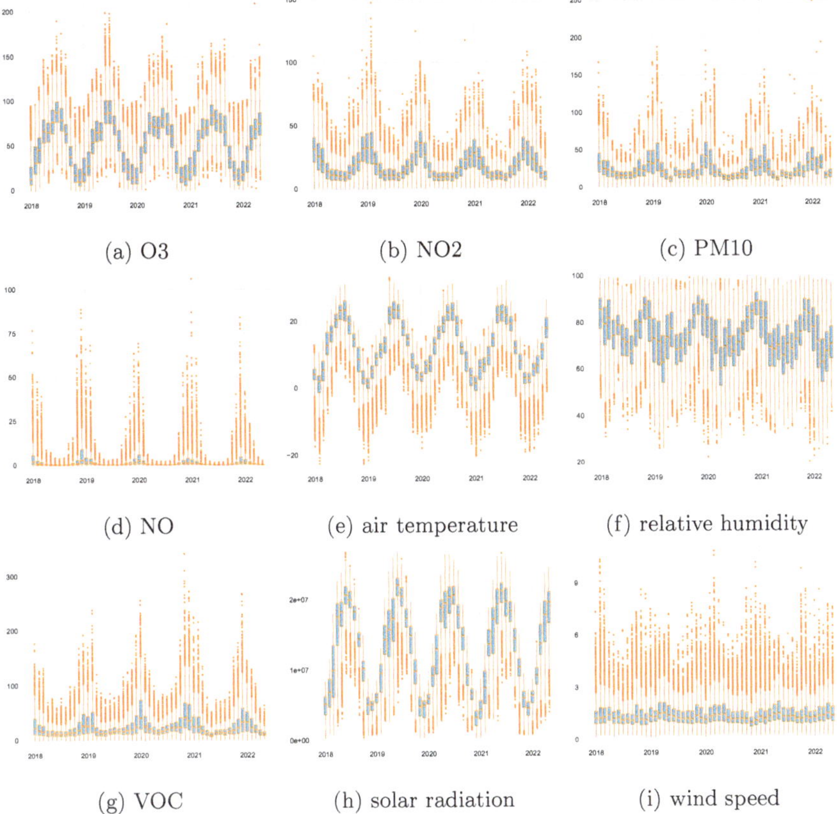

Fig. 1 (**a**)–(**i**) Boxplots of in situ and satellite-derived data, grouped by month for January 2018–May 2023

- Three ball kernels for the temporal domain (radii of 1, 2, and 3 days)
- Three ball kernels for the spatial domain (radii of 10, 15, and 20 km)
- Nine ring kernels for the spatiotemporal domain, combining temporal lags (1, 2, and 3 days) with spatial lags (0–10 km), (10–15 km), and (15–20 km)

The selection of kernel parameters has been guided by the analysis of temporal lags showing strongest correlations and by descriptive statistics of the Euclidean distances between spatial locations.

In Table 1, the parameters of the chosen kernel functions and the corresponding pseudo-eigenvalues are summarized. These latter have allowed the identification of the most relevant kernel configurations for each latent component. Higher pseudo-eigenvalues for temporal lags 1, 2, and 3 are particularly evident for z_1, z_2, and z_3, indicating a stronger short-term temporal dependence for these components. Concerning the spatial domain, the lags corresponding to 10 km, 15 km, and 20 km exhibit high pseudo-eigenvalues across nearly all components,

Table 1 Kernel settings, in terms of type, spatial and temporal radius (r_s and r_t, respectively), and pseudo-eigenvalues for the 15 kernel functions

Kernel function	r_s (r_{s_0}, r_{s_1})	r_t	z_1	z_2	z_3	z_4	z_5	z_6	z_7	z_8	z_9
Ball	0	1	1.174	0.959	1.013	0.795	0.728	0.724	0.454	0.833	0.771
Ball	0	2	0.857	0.475	0.685	0.424	0.304	0.325	0.110	0.540	0.412
Ball	0	3	0.619	0.233	0.542	0.251	0.132	0.194	0.075	0.426	0.273
Ball	10	0	5.293	5.011	4.060	3.432	3.685	2.935	3.017	2.034	1.889
Ball	15	0	6.948	6.404	4.741	4.041	4.361	3.380	3.391	2.285	1.823
Ball	20	0	8.250	7.375	5.151	4.400	4.791	3.665	3.598	2.473	1.742
Ring	(0, 10)	1	6.265	4.890	4.383	3.057	2.756	2.089	1.053	1.820	1.426
Ring	(0, 10)	2	4.656	2.427	3.094	1.688	1.009	0.805	0.053	1.081	0.752
Ring	(0, 10)	3	3.372	1.112	2.459	0.927	0.248	0.356	0.054	0.770	0.403
Ring	(10, 15)	1	5.488	4.104	3.001	2.190	1.970	1.402	0.579	1.174	0.523
Ring	(10, 15)	2	4.094	2.083	2.204	1.280	0.728	0.512	−0.054	0.715	0.273
Ring	(10, 15)	3	2.958	0.943	1.774	0.732	0.143	0.181	0.001	0.516	0.118
Ring	(15, 20)	1	5.453	3.872	2.597	1.931	1.775	1.275	0.453	1.086	0.346
Ring	(15, 20)	2	4.070	2.001	1.939	1.155	0.641	0.461	−0.083	0.689	0.191
Ring	(15, 20)	3	2.934	0.898	1.573	0.664	0.085	0.147	−0.003	0.507	0.095

confirming the relevance of spatial interactions. The ring kernels, which combine spatial and temporal dependence structures, also show that the spatiotemporal lags (0, 10) and (10, 15) are especially influential for the first and second components. Overall, these results indicate that both short-term temporal and medium-range spatial dependencies play a dominant role in shaping the latent spatiotemporal dynamics captured by the model. As discussed in [31], the outputs obtained from the stSOBI procedure, i.e., the pseudo-eigenvalues, the unmixing matrix ($\mathbf{W}$), and the estimated latent processes ($\mathbf{z}$), can be further exploited for subsequent analyses.

- II. **Model Each Component Univariately**

  The main advantage of ST-BSS is that the recovered latent processes are spatiotemporally independent, simplifying the subsequent modeling step from a complex multivariate problem to several manageable univariate problems. The spatiotemporal covariances have been estimated for $K = 189$ spatiotemporal lags (i.e., 9 spatial lags and 21 temporal lags) for each of the nine latent components. Their empirical characteristics have been analyzed [15, 21, 25], and sample non-separability ratios [2, 5, 13, 24] have been calculated for each latent component; since these ratios are smaller than 1 for each spatial and temporal lags, all empirical covariance functions could be modeled by a uniformly negative non-separable space-time covariance function. Hence, the product-sum model, which is uniformly negative non-separable and consistent with the detected features [16, 21], has been considered:

Table 2 Estimates of the product-sum models' parameters in (1) for the independent components of the ST-BSS model

	z_1	z_2	z_3	z_4	z_5	z_6	z_7	z_8	z_9
k_1	0.12	0.31	0.38	0.53	0.52	0.54	0.72	0.63	0.83
k_2	0.85	0.67	0.43	0.44	0.42	0.37	0.20	0.20	0.07
k_3	0.03	0.02	0.19	0.03	0.06	0.09	0.08	0.17	0.1
a_s	65 km	100 km	27 km	23 km	36 km	40 km	75 km	10 km	15 km
a_t	3.5 days	1.8 days	2 days	1.5 days	1.15 days	1 days	1 days	1.6 days	1.2 days

$$c_i(\mathbf{h}_s, h_t) = k_{1_i} C_{s_i}(\mathbf{h}_s) C_{t_i}(h_t) + k_{2_i} C_{s_i}(\mathbf{h}_s) + k_{3_i} C_{t_i}(h_t), \qquad i = 1, \ldots, 9, \tag{1}$$

with $k_{1_i} > 0$, $k_{2_i} \geq 0$, $k_{3_i} \geq 0$ and exponential covariance models both in space and in time, $C_{s_i} = Exp(||\mathbf{h}_s||; a_{s_i})$ and $C_{t_i} = Exp(|h_t|; a_{t_i})$, $i = 1, \ldots, 9$. The models' parameters (i.e., spatial range a_s, temporal range a_t, and coefficients k_1, k_2, k_3) have been estimated through nonlinear regression and detailed in Table 2.

- III. **Back-Transformation and Prediction Performance**
  The adequacy of the models derived through the ST-BSS approach has been assessed by performing a leave-one-out cross-validation for each latent component. Once the adequacy of the models had been verified, the matrix of the predicted latent components has been multiplied by the inverse of the unmixing matrix to reconstruct the target variable. Then, the O3 reconstructed values have been re-scaled using their corresponding mean and standard deviation, and the previously estimated seasonal component has been added to obtain the final O3 predictions. The comparison between observed and predicted O3 values has yielded a correlation coefficient of $r = 0.991$, a mean error (ME) of -0.05 and a mean absolute error (MAE) of 0.11. These results have confirmed the high accuracy and robustness of the models identified through the ST-BSS methodology.

4 Spatiotemporal Ozone Predictions

The models previously detected have been used for predicting the latent components across the study area and for time periods extending beyond the available observations. The resulting predictions, as described in Sect. 3.1, have been multiplied by the inverse of the unmixing matrix, re-scaled, and subsequently combined with the previously estimated seasonal component to obtain the daily O3 predictions for the three consecutive days following the last available observation, i.e., May 31, 2023. Figure 2 displays color maps illustrating the spatial distribution of the ozone levels on a grid (72×33), with a very high spatial resolution.

It is evident that during the first 3 days of June 2023, ground-level ozone concentrations over Northern Italy exhibited elevated values, particularly across the

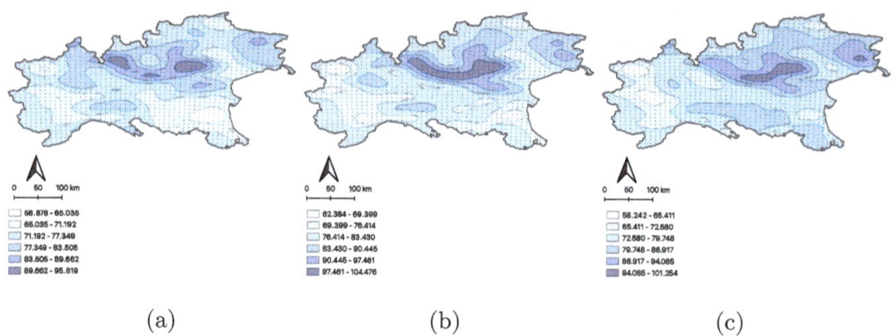

Fig. 2 Color maps showing O3 predictions in. (**a**) June 1, 2023. (**b**) June 2, 2023. (**c**) June 3, 2023

Po Valley, where stagnant meteorological conditions, high solar radiation, and rising temperatures favored intense photochemical activity. Such conditions are consistent with early-summer ozone episodes typically observed in this region, driven by the accumulation of precursors and limited atmospheric dispersion. These elevated levels are of immediate relevance not only for public health warnings but also for supporting policy makers in activating alert thresholds and mitigation measures, such as temporary traffic restrictions, enhancement of public transport services, or health notice to vulnerable groups. It is worth noting that, from a methodological point of view, the application of the ST-BSS procedure has substantially reduced computational time and model complexity, since it is not necessary to estimate and model cross-dependencies among variables.

5 Conclusions

Modeling multivariate spatiotemporal data, while inherently complex, is crucial for understanding and predicting complex phenomena, such as ozone behavior. The ST-BSS based approach offered powerful methodology to address this challenge by transforming complex multivariate dependencies into simpler univariate latent components. Future research in this domain will continue to explore various extensions, including the development of deep learning methods that are robust to outliers and tackle the problem of huge amount of missing values.

Bibliography

1. Cabral Pinto, F., Manchuk, J.G., Deutsch, C.V.: Decomposition of multivariate spatial data into latent factors. Comput. Geosci. **153**, 104773 (2021)
2. Cappello, C., De Iaco, S., Posa, D.: Testing the type of non-separability and some classes of space-time covariance function models. Stochastic Environ. Res. Risk Assess. **32**, 17–35 (2018)

3. Cappello, C., De Iaco, S., Maggio, S., Posa, D.: Modeling ocean currents through complex random fields indexed in time. Math. Geosci. **53**, 999–1025 (2020)
4. Cappello, C., De Iaco, S., Maggio, S., Posa, D.: Time varying complex covariance functions for oceanographic data. Spatial Stat. **42**(4), 100426 (2020)
5. Cappello, C., De Iaco, S., Posa, D.: covatest: an R package for selecting a class of space-time covariance functions. J. Stat. Softw. **94**(1), 1–42 (2020)
6. Cappello, C., De Iaco, S., Palma, M., Pellegrino, D.: Spatio-temporal modeling of an environmental trivariate vector combining air and soil measurements from Ireland. Spatial Stat. **42**, 100455 (2021)
7. Cappello, C., De Iaco, S., Palma, M.: Computational advances for spatio-temporal multivariate environmental models. Comput. Stat. **37**(2), 651–670 (2022)
8. Cappello, C., De Iaco, S., Palma, M., Nordhausen, K.: Space-time blind source separation and linear coregionalization modeling: a critical comparison. Stochastic Environ. Res. Risk Assess. **39**, 6373–6397 (2025)
9. De Iaco, S.: The cgeostat software for analyzing complex-valued random fields. J. Stat. Softw. **79**(5), 1–32 (2017)
10. De Iaco, S.: New spatio-temporal complex covariance functions for vectorial data through positive mixtures. Stochastic Environ. Res. Risk Assess. **36**, 2769–2787 (2022)
11. De Iaco, S.: Families of complex-valued covariance models through integration. Environmetrics **34**(3), 1–16 (2023)
12. De Iaco, S.: Spatio-temporal generalized complex covariance models based on convolution. Comput. Stat. Data Anal. **183**, 107709 (2023)
13. De Iaco, S., Posa, D.: Positive and negative non-separability for space-time covariance models. J. Stat. Plan. Inference **143**(2), 378–391 (2013)
14. De Iaco, S., Posa, D.: Wind velocity prediction through complex kriging: formalism and computational aspects. Environ. Ecol. Stat. **23**(1), 115–139 (2016)
15. De Iaco, S., Posa, D.: Strict positive definiteness in geostatistics. Stochastic Environ. Res. Risk Assess. **32**, 577–590 (2018)
16. De Iaco, S., Myers, D.E., Posa, D.: Space-time analysis using a general product-sum model. Stat. Probab. Lett. **52**(1), 21–28 (2001)
17. De Iaco, S., Myers, D.E., Posa, D.: Space-time variograms and a functional form for total air pollution measurements. Comput. Stat. Data Anal. **41**(2), 311–328 (2002)
18. De Iaco, S., Myers, D.E., Posa, D.: The linear coregionalization model and the product-sum space-time variogram. Math. Geol. **35**(1), 25–38 (2003)
19. De Iaco, S., Palma, M., Posa, D.: Covariance functions and models for complex-valued random fields. Stochastic Environ. Res. Risk Assess. **17**(3), 145–156 (2003)
20. De Iaco, S., Palma, M., Posa, D.: Modeling and prediction of multivariate space-time random fields. Comput. Stat. Data Anal. **48**(3), 525–547 (2005)
21. De Iaco, S., Posa, D., Myers, D.E.: Characteristics of some classes of space-time covariance functions. J. Stat. Plan. Infer. **143**(11), 2002–2015 (2013)
22. De Iaco, S., Posa, D., Palma, M.: Complex-valued random fields for vectorial data: estimating and modeling aspects. Math. Geosci. **45**(5), 557–573 (2013)
23. De Iaco, S., Palma, M., Posa, D.: Choosing suitable linear coregionalization models for spatio-temporal data. Stochastic Environ. Res. Risk Assess. **33**(7), 1419–1434 (2019)
24. De Iaco, S., Posa, D., Cappello, C., Maggio, S.: Isotropy, symmetry, separability and strict positive definiteness for covariance functions: a critical review. Spat. Stat. **29**, 89–108 (2019)
25. De Iaco, S., Posa, D., Cappello, C., Maggio, S.: On some characteristics of Gaussian covariance functions. Int. Stat. Rev. **89**(1), 36–53 (2021)
26. De Iaco, S., Cappello, C., Palma, M., Nordhausen, K.: A multivariate approach for modeling spatio-temporal agrometeorological variables. Environmetrics **36**(2), e2891 (2025)
27. Genton, M., Kleiber, W.: Cross-covariance functions for multivariate geostatistics. Stat. Sci. **30**(2), 147–163 (2015)
28. Grzebyk, M.: Ajustement d'une Coregionalisation Stationnaire. Ph.D. thesis, Ecoles des Mines, Paris (1993)

29. Lajaunie, C., Béjaoui, R.: Sur le krigeage des fonctions complexes. Technical Report, N-23/91/G, Centre de Geostatistique, Ecole des Mines de Paris, Fontainebleau (1991)
30. Muehlmann, C., Nordhausen, K., Yi, M.: On cokriging, neural networks, and spatial blind source separation for multivariate spatial prediction. IEEE Geosci. Remote Sens. Lett. **18**(11), 1931–1935 (2021)
31. Muehlmann, C., De Iaco, S., Nordhausen, K.: Blind recovery of sources for multivariate space-time random fields. Stoch. Environ. Res. Risk Assess. **37**, 1593–1613 (2023)
32. Muehlmann, C., Piccolotto, N., Cappello, C., De Iaco, S., Nordhausen, K.: SpaceTimeBSS: Blind Source Separation for Multivariate Spatio-Temporal Data (2023), https://CRAN.R-project.org/package=SpaceTimeBSS, R package version 0.3-0
33. Muehlmann, C., Filzmoser, P., Nordhausen, K.: Spatial blind source separation in the presence of a drift. Aust. J. Stat. **53**(2), 48–68 (2024)
34. Muehlmann, C., Cappello, C., De Iaco, S., Nordhausen, K.: Anisotropic local covariance matrices for spatial blind source separation. AStA Adv. Stat. Anal. **109**, 753–770 (2025)
35. Nordhausen, K., Oja, H., Filzmoser, P., Reimann, C.: Blind source separation for spatial compositional data. Math. Geosci. **47**(7), 753–770 (2015)
36. Nuvolone, D., Petri, D., Voller, F.: The effects of ozone on human health. Environ. Sci. Pollut. Res. **25**(9), 8074–8088 (2018)
37. Posa, D.: Parametric families for complex valued covariance functions: some results, an overview and critical aspects. Spat. Stat. **39**, 1–20 (2020)
38. Posa, D.: Models for the difference of continuous covariance functions. Stoch. Environ. Res. Risk Assess. **35**, 1369–1386 (2021)
39. Rouhani, S., Wackernagel, H.: Multivariate geostatistical approach to space-time data analysis. Water Resour. Res. **26**(4), 585–591 (1990)
40. Sipilä, M., Cappello, C., De Iaco, S., Nordhausen, K., Taskinen, S.: Modelling multivariate spatio-temporal data with identifiable variational autoencoders. Neural Netw. **181**, 106774 (2025)
41. Wackernagel, H.: Multivariate Geostatistics: an Introduction with Applications. Springer, Berlin (2003)
42. Yaglom, A.M.: Correlation theory of stationary and related random functions. Springer Series in Statistics, vols. I, II. Springer, Berlin (1987)
43. Zhang, J., Wei, Y., Fang, Z.: Ozone pollution: a major health hazard worldwide. Front. Immunol. **10**, 2518 (2019)

Zeitfracht Medien GmbH
Ferdinand-Jühlke-Straße 7
99095 Erfurt, Deutschland
produktsicherheit@kolibri360.de